Selected Titles in This Series

4 **Robert Friedman and John W. Morgan, Editors,** Gauge Theory and the Topology of Four-Manifolds, 1998

3 **János Kollár, Editor,** Complex Algebraic Geometry, 1997

2 **Robert Hardt and Michael Wolf, Editors,** Nonlinear Partial Differential Equations in Differential Geometry, 1996

1 **Daniel S. Freed and Karen K. Uhlenbeck, Editors,** Geometry and Quantum Field Theory, 1995

Ω IAS/PARK CITY
MATHEMATICS SERIES

Volume 4

Gauge Theory and the Topology of Four-Manifolds

Robert Friedman
John W. Morgan
Editors

American Mathematical Society
Institute for Advanced Study

IAS/Park City Mathematics Institute runs mathematics education programs that bring together high school mathematics teachers, researchers, graduate students, and undergraduates to participate in distinct but overlapping programs of research and education. This volume contains the lecture notes from the Graduate Summer School program on gauge theory and smooth 4-manifolds held in Park City, Utah, on July 10–30, 1994.

Supported by the National Science Foundation.

1991 *Mathematics Subject Classification.* Primary 14–XX, 32–XX, 53–XX, 57–XX, 58–XX.

Library of Congress Cataloging-in-Publication Data

Gauge theory and the topology of four-manifolds / Robert Friedman, John W. Morgan, editors.
p. cm. — (IAS/Park City mathematics series, ISSN 1079-5634 : v. 4)
Includes bibliographical references.
ISBN 0-8218-0591-6 (alk. paper)
1. Four-manifolds (Topology) 2. Gauge fields (Physics) 3. Mathematical physics. I. Friedman, Robert, 1955– . II. Morgan, John W., 1946– . III. Series.
QC20.7.M24G38 1997
530.14′35—dc21 97-32356
CIP

♾ The paper used in this book is acid-free and falls within the guidelines
established to ensure permanence and durability.
Visit the AMS home page at URL: `http://www.ams.org/`

10 9 8 7 6 5 4 3 2 1 03 02 01 00 99 98

Contents

Preface

The IAS/Park City Mathematics Institute (PCMI) was founded in 1991 as part of the "Regional Geometry Institute" initiative of the National Science Foundation. In mid 1993 the program found an institutional home at the Institute for Advanced Study (IAS) in Princeton, New Jersey. The PCMI will continue to hold summer programs alternately in Park City and in Princeton.

The IAS/Park City Mathematics Institute encourages both research and education in mathematics and fosters interaction between the two. The three-week summer institute offers programs for researchers and postdoctoral scholars, graduate students, undergraduate students, high school teachers and undergraduate faculty. One of PCMI's main goals is to make all of the participants aware of the total spectrum of activities that occur in mathematics education and research: we wish to involve professional mathematicians in education and to bring modern concepts in mathematics to the attention of educators. To that end the summer institute features general sessions designed to encourage interaction among the various groups. Beginning in 1997 the summer institute also includes an Undergraduate Faculty Program. In-year activities at sites around the country form an integral part of the Program for High School Teachers.

Each summer a different topic is chosen as the focus of the Research Program and Graduate Summer School. Activities in the Undergraduate Program deal with this topic as well. Lecture notes from the Graduate Summer School are being published each year in this series. The first four volumes are:

Volume 1: *Geometry and Quantum Field Theory* (1991)
Volume 2: *Nonlinear Partial Differential Equations in Differential Geometry* (1992)
Volume 3: *Complex Algebraic Geometry* (1993)
Volume 4: *Gauge Theory and the Topology of Four-Manifolds* (1994)

Future volumes from the 1995 Summer School *Nonlinear Wave Phenomena*, the 1996 Summer School *Probability*, and the 1997 Summer School *Symplectic Geometry and Topology* are in preparation. The 1998 Research Program and Graduate Summer School topic is *Representation Theory of Lie Groups* and the 1999 Research Program and Graduate Summer School topic is *Number Theory*.

We plan to publish material from other parts of the IAS/Park City Mathematics Institute in the future. This will include material from the Undergraduate Program, the Program for High School Teachers and publications documenting the interactive activities which are a primary focus of the PCMI. At the summer institute late afternoons are devoted to seminars of common interest to all participants. Many deal with current issues in education; others treat mathematical topics at a level which encourages broad participation. The PCMI has also spawned interactions between universities and high schools at a local level. We hope to share these activities with a wider audience in future volumes.

Dan Freed, Series Editor
September, 1997

Introduction

The 1994 Summer School in Gauge Theory and Smooth 4-manifolds occurred at a unique moment in the history of the subject. The theory had been developing and maturing for about a dozen years, following the basic outlines laid down by Donaldson in his seminal work on the applications of gauge theory to 4-manifold topology. Many basic questions that seemed accessible by gauge theory techniques had been set forth and, to varying degrees, many partial answers to these questions were known. New results at the time included an evolving understanding of the overall algebraic structure of the invariants including a complete blow-up formula and the notion of simple type 4-manifolds and their basic classes. New and powerful C^∞ techniques for computing Donaldson invariants had been introduced, replacing difficult arguments in algebraic geometry with gluing formulas and cut-and-paste methods. These techniques gave the first complete calculations for Donaldson invariants of elliptic surfaces, as well as a way to show that certain simply connected 4-manifolds could not be written as a connected sum of algebraic surfaces. The fact that the basic classes of many Kähler manifolds are all multiples of the canonical class had been established. There was also a "physics proof" that this was true in general. Using algebraic geometry, the smooth invariance of Kodaira dimension of an algebraic surface had been established, but deeper questions involving the canonical class, the exceptional curves, and the plurigenera of Kähler manifolds did not seem accessible. Progress had been spectacular, but was slowing, and the remaining open questions seemed to involve formidable technical difficulties.

At almost exactly the same time that the summer school took place, a preprint by Seiberg-Witten appeared on the physics preprint network. The paper concerned electro-magnetic duality in gauge theories. It went totally unnoticed in the 4-manifold community until the fall of 1994, when Witten showed how to use the ideas of that paper to compute Donaldson invariants of a 4-manifold in terms of certain $U(1)$-invariants, now called the Seiberg-Witten invariants. This paper definitely was noticed by the mathematical community, where it caused an overnight revolution unparalleled in our mathematical experience. Within weeks most long-standing conjectures from Donaldson theory were established, one after another, and the proofs of the known results were simplified dramatically. The understanding of the invariants for a Kähler manifold was completed, as well as the consequence that for all Kähler manifolds except for rational and ruled surfaces, the canonical class of the minimal model and the classes of the exceptional curves are differentiable invariants. Also, the long-standing Thom conjecture about the genus of a Riemann surface in the complex projective plane was resolved in the affirmative. Finally, there was the revelation that much of the classification theory for Kähler manifolds generalizes to symplectic manifolds. These are all developments that lay in the future when this summer school was in progress. They do not appear here. Surveys[1] of Seiberg-Witten theory and its relation to Donaldson theory are now available. A detailed exposition[2] of Seiberg-Witten theory is also available.

What then is the raison d'être for this volume? We see several. First, the lectures here provide a historical perspective akin to the movies of workers programming the UNIVAC computer by connecting plugs. That is to say, one can view this volume as showing how the ancients studied 4-manifold theory before the invention of the modern-day Seiberg-Witten theory. In a more serious vein, one reason that the progress using the Seiberg-Witten invariants was so spectacular was that those studying $SU(2)$-gauge theory had more than ten years' experience with the subject. The tools had been honed, the correct questions formulated, the basic strategies and approaches were by then well-understood. This knowledge immediately bore fruit in the technically simpler environment of the Seiberg-Witten theory. But surely the progress would not have been so rapid without the ten years of struggle with $SU(2)$-gauge theory as background. For this reason we think it is worthwhile to have an account of this version of the subject. Lastly, as far as the polynomial invariants are concerned there is no reason to prefer the more complicated $SU(2)$-theory to the simpler Seiberg-Witten $U(1)$-theory. But the $SU(2)$-moduli spaces are much richer geometrically and much more closely tied in with the geometry of the underlying 4-manifold. By contrast, the Seiberg-Witten moduli spaces tend to be finite sets of points and hence contain no more information than the algebraic invariants they define. Indeed, gauge theory has a long prehistory predating Donaldson's applications of the subject to 4-manifold topology, where the central concerns of the subject were the overall study of the geometry of the moduli space. One reason for the interest in this study is the connection between the gauge theory moduli spaces of a Kähler manifold and the algebro-geometric moduli space of stable holomorphic bundles over the manifold. One can easily imagine that the time will come when the extra geometric richness of the $SU(2)$-moduli spaces will be important for purposes beyond the algebraic invariants that have been studied to date. At that point the techniques and results presented here will be essential.

The first two papers in this volume concern the theory of stable holomorphic bundles over algebraic surfaces. The article by Gieseker describes his construction via geometric invariant theory of the compactification of the moduli space of stable bundles over a surface. The article by Jun Li shows, in the case of a Kähler manifold base, that every stable holomorphic bundle admits a unique anti-self-dual connection. Together with the much easier converse, which says that an irreducible anti-self-dual connection on a principal $U(n)$-bundle defines a holomorphic structure on the associated n-plane bundle which is stable with respect to the Kähler form, this shows that there is an isomorphism between the moduli space of stable holomorphic bundles and the moduli space of anti-self-dual connections. This isomorphism implies that one can study the Donaldson polynomial invariants of Kähler manifolds using the techniques of stable bundle theory. Many explicit computations of the invariants of algebraic surfaces have been done in this way. More generally, the techniques of algebraic geometry are then available for giving a concrete description of the anti-self-dual moduli space.

[1]S.K. Donaldson, *The Seiberg-Witten equations and 4-manifold topology*, Bull. Amer. Math. Soc. (N.S.), **33** (1996), 45–71 and R. Friedman, *Donaldson and Seiberg-Witten invariants of algebraic surfaces*, Algebraic Geometry (Santa Cruz 1995), Proc. Symp. Pure Math., American Mathematical Society, Providence (to appear)

[2]J. W. Morgan, *The Seiberg-Witten Equations and Applications to the Topology of Smooth Four-Manifolds*, Mathematical Notes, vol 44, Princeton University Press, Princeton, 1996

The goal of the next article by Morgan is to give the definition of the Donaldson polynomial invariants and to state two basic results in the subject – a non-vanishing theorem for the invariants of an algebraic surface (which depends on the results of Jun Li's lecture) and a vanishing theorem for connected sums. As a consequence of these two results, one is able to derive Donaldson's result that algebraic surfaces are not connected sums in a nontrivial way. The first four lectures lay out the basic foundational material of gauge theory – principal bundles, connections, covariant differentiation, curvature, the infinite dimensional manifold of equivalence classes of connections. This elementary material is explained in detail and the exposition is aimed at a student who has a basic knowledge of differentiable manifolds and differential forms. The last four lectures study the anti-self-dual equations and establish the basic properties of the moduli space of solutions of these equations. Then there is a brief introduction to some of the main analytic tools needed to understand the nature of infinity in the moduli space as well as establish the gluing theorem and hence the vanishing theorem for connected sums, namely Uhlenbeck's compactness theorem and Taubes' gluing result. The article then sketches the definitions of Donaldson's polynomial invariants and finishes with the two results indicated above. This material is more advanced and the purpose of the lectures is to give an overview of the basic results and techniques rather than a complete account, such as can be found in other references.[3]

The next article, by Stern, gives an overview of the state of the subject of 4-manifold theory. He lists important questions in the subject as well as various results obtained from gauge theory techniques. After this introduction, the body of the article concerns explicit computations of the invariants in various situations. The basic technique explained in this article is to use cut-and-paste formulae, arising out of the gluing theorems along rational homology spheres, to compute relationships between invariants of related manifolds. Given enough of these relationships, one can explicitly solve for the invariants. One of the nicest examples of this method is the explicit blow-up formula, expressing the invariants of a connected sum of a manifold with the complex projective plane in terms of the invariants of the original manifold. Stern gives other examples of this method, including a derivation of the form of the invariants for a manifold of simple type and a complete determination of the invariants of elliptic surfaces. This article indicates how far one can get with some of the easier gluing theorems and clever topological decompositions.

The last article, written by Bryan based on lectures of Taubes, lays down a formalism for studying general gluing theorems and product formulae. The formalism provides a context for extending the product formulae from Donaldson-Floer theory, for decompositions along homology three-spheres, and the product formulae for decompositions along rational homology three-spheres, used so brilliantly by Fintushel and Stern, to product formulae for decompositions along arbitrary closed 3-manifolds. Such a general product formula will be a powerful tool in understanding the nature of the invariants on a general 4-manifold.

In addition to the lecture series represented by articles that appear here, there were two other series. One series by Bob Friedman on stable holomorphic vector bundles and the other by Tom Mrowka on analytic and topological aspects on the

[3]S.K. Donaldson and P. Kronheimer, *The Geometry of Four-Manifolds*, Clarendon Press, Oxford, 1990 and R. Friedman and J. W. Morgan, *Smooth Four-Manifolds and Complex Surfaces*, Ergebnisse der Mathematik und ihrer Grenzgebiete 3. Folge **27**, Springer, Berlin Heidelberg New York, 1994

Donaldson invariants. A written version of Friedman's lectures appear independently[4].

Many people helped contribute to the success of the 1994 Summer School. It is a pleasure to thank all the lecturers who worked hard to give clear and accessible presentations and to thank all the teaching assistants who helped prepare the manuscripts and lead the problem sessions. We would also like to thank all the students and researchers whose enthusiastic participation in the summer school made it the lively success that it was. On the administrative side, the tireless work of Anne Humes and Lorena Hitchens was absolutely essential to the program. It is a pleasure to thank them for their heroic efforts which were always accompanied by unfailing good humor. We also greatly appreciated the advice and help of Herb Clemens, Dan Freed and Karen Uhlenbeck as we were putting together this program. Their years of experience with the summer school were of great benefit to us. Lastly, it is a pleasure to thank John Polking for his unstinting leadership of IAS/PCMI. Without this leadership there would be no IAS/PCMI.

John Morgan and Robert Friedman, Volume Editors
September, 1997

[4]Algebraic Surfaces and Holomorphic Vector bundles, Springer-Verlag, 1998

Geometric Invariant Theory and the Moduli of Bundles

David Gieseker

IAS/Park City Mathematics Series
Volume 4, 1998

Geometric Invariant Theory and the Moduli of Bundles

David Gieseker

LECTURE 1
Geometric Invariant Theory

Let $G = SL(n, \mathbf{C})$ and suppose that G operates on a variety X. We consider the set of orbits X/G. Geometric invariant theory addresses the following question: When can we define X/G as a variety Y, i.e., the points of X/G correspond in a natural way to the points of Y? The first problem we encounter is that X/G with the quotient topology is usually not Hausdorff. This can be seen in the following example:

Example. Given an integer N, let V_N be the set of all homogeneous polynomials of degree N in two variables X_0 and X_1. Let $\mathbf{P}(V_N)$ be the set of lines in V_N. $G = SL(2, \mathbf{C})$ operates on V_N in the following way: If $g = (a_{ij}) \in G$ and $P \in V_N$, then

$$P^g(X_0, X_1) = P(g^{-1} \begin{pmatrix} X_0 \\ X_1 \end{pmatrix}).$$

We have the following geometric interpretation of $f \in V_N$. The equation $\{f = 0\}$ defines a finite set of points with multiplicities in $\mathbf{P}^1$. The multiplicity of $p \in \mathbf{P}^1$ is just the order of vanishing of f at p. Thus $\{f = 0\}$ defines a divisor D_f on $\mathbf{P}^1$, and $\mathbf{P}(V_N)$ exactly corresponds to the space of divisors of degree N on $\mathbf{P}^1$. The corresponding action of G on the space of divisors is to move the divisors around by linear fractional transformations. We cannot form X/G as a variety ($X = \mathbf{P}(V_N)$), since X/G is not Hausdorff in the classical topology. Indeed, consider $\bar{f} \in \mathbf{P}(V_N)$ and $f \in V_N$ a corresponding equation. We can find a g in the orbit of f so that X_1^N occurs in $g(X_0, X_1)$ with coefficient one. Set $g_t(X_0, X_1) = t^N g(tX_0, t^{-1}X_1)$. Then we cannot give X/G a Hausdorff structure so that $\phi : X \to X/G$ is continuous: If ϕ were continuous, from

$$\lim_{t \to 0} g_t(X_0, X_1) = g_0(X_0, X_1) = X_1^N,$$

[1]Department of Mathematics, UCLA
E-mail address: dag@math.ucla.edu

we would have $\phi(f) = \phi(g) = \lim_{t\to 0}\phi(g_t) = \phi(X_1^N)$. So ϕ would have to map all orbits onto one element.

In the general case, consider $G = SL(n, \mathbf{C})$ operating on a vector space V, i.e., there is an algebraic map $\phi : G \to GL(V)$ which is a homomorphism. We will sometimes denote $\phi(g)(v)$ by v^g for $g \in G$ and $v \in V$. The idea is to find a large open subset $U_s \subseteq \mathbf{P}(V)$ so that U_s/G can be given the structure of a variety. To construct U_s/G, we look at the space $W_N = S^N(V^*)$ of homogeneous polynomials of large degree N on V. G operates on W_N by

$$P^g(v) = P(g^{-1}(v)).$$

We say P is invariant if $P^g = P$ for all $g \in G$. Let $P_0, ..., P_k$ be a basis for the space of all invariant polynomials in W_N. We attempt to use $P_0, ..., P_k$ to define a map

$$\Phi_N : \mathbf{P}(V) \dashrightarrow \mathbf{P}^k$$

by $\Phi_N(v) = (P_0(v), ..., P_k(v))$. The problem with this map is that $P_0(v), ..., P_k(v)$ could be all zero, so that they would not define a point in $\mathbf{P}^k$. This can happen if $0 \in \overline{v^G}$, where v^G is the orbit of v. For any homogeneous polynomial of positive degree P vanishes at 0 and is constant on $\overline{v^G}$. So P vanishes at v.

Definition:.
i) v is semistable if $0 \notin \overline{v^G}$;
ii) v is stable if v^G is closed and v has finite stabilizer;
iii) v is weakly stable if v^G is closed.

A point $\overline{v} \in \mathbf{P}(V)$ is stable (resp., semistable, weakly stable) if any lift $v \in V$ of $\overline{v}$ is stable (resp., semistable, weakly stable). Let U_s (resp., $U_{s.s.}$) be the set of stable (resp., semistable) points in $\mathbf{P}(V)$.

Theorem. *If N is sufficiently large, Φ_N is defined exactly on $U_{s.s.}$. The map from U_s/G to $\mathbf{P}^k$ is injective.*

Proof. We shall use Weyl's unitary trick. Let $U \subset G = SL(n, \mathbf{C})$ be the subgroup of special unitary matrices. Then U is Zariski dense in G. In particular, if P is any algebraic function which vanishes on U, then P vanishes on G. To see that U is Zariski dense in G, note that the Zariski closure of U is an algebraic group whose tangent space at I includes the tangent space of U at I. But the tangent space T of U at I is the space of traceless matrices A with $A = \bar{A}^t$. The complex span of T is all of the traceless matrices, so $\overline{U} = G$.

There is a U invariant measure μ on U with

$$\int_U d\mu = 1.$$

We define for $P \in V_N$,

$$R(P) = \int_{g\in U} P^{g^{-1}}(v)d\mu.$$

Note $R(P) \in V_N$ and

$$R(P)(v) = \int_{g\in U} P(vg)d\mu.$$

Suppose X_1 and X_2 are two disjoint Zariski closed G invariant subsets of V. We will now show that there is an invariant $P \in V_N$ so that P is 1 on X_1 and 0 on X_2. Let I_i be the ideal of X_i. Then the functions in the ideal $I_1 + I_2$ have no common zeroes. So $1 \in I_1 + I_2$ by Hilbert's Nullstellensatz. (In fact, this is the original application of the Nullstellensatz.) Hence $1 = f_1 + f_2$ for some $f_i \in I_i$, and f_2 is identically 1 on X_1 and identically 0 on X_2. Let $P = R(f_2)$. Then P is G invariant and is 1 on X_1 and 0 on X_2, since $1 = R(f_1) + R(f_2)$. If v is semistable, set $X_1 = \overline{v^G}$ and $X_2 = \{0\}$. There is an invariant polynomial P which is 1 on X_1 and 0 on X_2. Write $P = \sum P_n$, where P_n is homogeneous of degree $n > 0$. Then all the P_n are invariant, and one must be nonzero at v.

Thus Φ_N is defined on any fixed semistable point if $N \gg 0$. We have to check that we can choose one N which works for all semistable points. We show this by verifying that the ring of invariant polynomials is finitely generated as a $\mathbf{C}$-algebra.

First note that if g is invariant, then $R(gf) = gR(f)$. Now let S be the ring of all invariant polynomials on V and let S^+ be the ones of positive degree. $\mathbf{C}[V] \cdot S^+$ is an ideal in $\mathbf{C}[V]$ and so finitely generated. In fact, we can take the generators to be of the form $h_i = f_i(X)g_i(X)$, where $g_i \in S^+$ and f_i homogeneous. If h is any element of S^+, then $h = \sum f_i g_i$, with $\deg(f_i) < \deg(h)$. So $h = R(h) = \sum R(f_i) g_i$. This shows, by induction on $\deg(h)$, that S is finitely generated as a $\mathbf{C}$-algebra.

There is a natural G invariant surjective map $\phi : \mathbf{P}(V)_{s.s.} \to \mathrm{Proj}(R^G)$. The induced map from the set of weakly stable orbits in $\mathbf{P}(V)_{s.s.}$ to $\mathrm{Proj}(R^G)$ is bijective.

The semistable replacement property:

Let C be a smooth algebraic curve and $P \in C$. Suppose $\phi : C - P \to V_{s.s.}$. Then there is a smooth curve C', a map $f : C' \to C$, a point $P' \in C'$ with $f(P') = P$ and a map $\sigma : C' - P' \to G$ so that if we define $\phi' : C' - P' \to V_{s.s.}$ by

$$\phi'(Q') = \phi(f(Q'))^{\sigma(Q')},$$

then ϕ' extends to a map $\phi' : C' \to V_{s.s.}$.

The existence of $\phi : \mathbf{P}(V)_{s.s.} \to \mathrm{Proj}(R^G)$ and the proof of the semi-stable replacement property are long. Of course, the usual reference is [**MFK**], but [**N**] provides a more elementary introduction to the subject.

LECTURE 2
The Numerical Criterion

How can we tell if a given $v \in V$ is semistable or not? Calculating the invariants is prohibitively difficult, even in our example of V_N. We have the following definition to help us:

Definition. *A one parameter subgroup* $(1-P-S)$ *of* G *is just a nontrivial algebraic homomorphism* $\lambda : \mathbf{C}^* \to G = SL(n, \mathbf{C})$.

It follows from elementary representation theory that there is a basis $v_1, ..., v_n$ of $\mathbf{C}^n$ and $r_i \in \mathbf{Z}$ so that $\lambda(\alpha)$ is diagonal with

$$\lambda(\alpha) = \begin{pmatrix} \alpha^{r_1} & & & \\ & \alpha^{r_2} & & \\ & & \cdots & \\ & & & \alpha^{r_n} \end{pmatrix}.$$

Note that we do not assume that λ is injective, so the terminology of subgroup is misleading. Actually, the $1-P-S$ we will need will already be in diagonal form, so we could take the diagonal form as a definition.

Given a $1-P-S$ $\lambda : \mathbf{C}^* \to V$, a G operation on V and $v \in V$, we can define $\phi : \mathbf{C}^* \to V$ by $\phi(\alpha) = v^{\lambda(\alpha)}$. We say $\lim_{\alpha \to 0} v^{\lambda(\alpha)} = \infty$ if ϕ cannot be extended to a map $\phi_1 : \mathbf{C} \to V$. If ϕ can be extended, we write $\lim_{\alpha \to 0} v^{\lambda(\alpha)} = \phi_1(0)$.

Theorem (The numerical criterion).
i) $\lim_{\alpha \to 0} v^{\lambda(\alpha)} = \infty$ *for all* $1-P-S$ λ *if and only if* v *is stable;*
ii) $\lim_{\alpha \to 0} v^{\lambda(\alpha)} \neq 0$ *for all* $1-P-S$ λ *if and only if* v *is semistable.*

Th proof uses the following elementary fact from algebraic geometry: If $f : X \to Y$ is a map of algebraic varieties and $y \in \overline{f(X)}$, then there is a curve C, a point $P \in C$ and maps

$$\begin{array}{ccc} C - P & \xrightarrow{\alpha} & X \\ \cap & & \downarrow f \\ C & \xrightarrow{h} & Y \end{array}$$

so that $h(P) = y$. That is, roughly speaking, we can approach y along a curve in $f(X)$. Let's suppose that v is not semistable. Applying the above remark with $Y = V$ and $X = G$, we can find $\mu : C - P \to G$ so that

$$\lim_{Z \to P} v^{\mu(Z)} = 0.$$

Let $\mathcal{O}$ be the local ring of P at C and K be the quotient field of $\mathcal{O}$. So in the analytic category, $\mathcal{O}$ is the ring of analytic functions in a neighborhood of P and K is the ring of Laurent series with finite tails. We can think of μ as defining an element $\mu \in SL(n, K)$, i.e., $\mu = (\mu_{ij})$ is a matrix of functions defined on a neighborhood of P. For t near P, $\mu(t) = (\mu_{ij}(t))$. Thus we can consider $\mu : K^n \to K^n$. Let t be a local parameter at P. Then t is, up to units, the unique irreducible element of $\mathcal{O}$ and $\mathcal{O}$ is a principal ideal domain. By the theory of modules over PID's, we can find σ_1 and σ_2 in $SL(n, \mathcal{O})$ so that

$$\sigma_1 \mu \sigma_2 = \begin{pmatrix} t^{r_1} & & & \\ & t^{r_2} & & \\ & & \dots & \\ & & & t^{r_n} \end{pmatrix} = \lambda(t)$$

by [**L**], Chapter III, Theorem 7.9.

We can assume that $r_1 \leq r_2 \leq \dots \leq r_n$. We claim that

$$\lim_{t \to 0} v^{\lambda(t)} = 0.$$

By changing basis in $\mathbf{C}^n$, we can assume that $\sigma_1(0) = I$. Let $e_1, \dots, e_n$ be the usual basis for $\mathbf{C}^n$ so that $e_i^{\lambda(t)} = t^{r_i} e_i$. Write $v = \sum x_j e_j$ and $e_i^{\sigma_1^{-1}(t)} = \sum a_{ij}(t) e_j$. Then

$$\begin{aligned} v^{\mu(t)\sigma_2(t)} &= \sum x_i e_i^{\mu(t)\sigma_2(t)} \\ &= \sum x_i e_i^{\sigma_1^{-1}(t)\lambda(t)} \\ &= \sum (\sum a_{ij}(t) x_i) e_j^{\lambda(t)} \\ &= \sum (\sum a_{ij}(t) x_i) t^{r_j} e_j \end{aligned}$$

Now $\lim_{t\to 0} v^{\mu(t)} = 0$, implying $\lim_{t \to 0} v^{\mu(t)\sigma_2(t)} = 0$. So $\lim_{t\to 0} \sum_i a_{ij}(t) x_i = \sum_i a_{ij}(0) x_i = x_j = 0$ if $r_j \leq 0$. Hence $\lim_{t \to 0} v^{\lambda(t)} = 0$. The criterion for stability is proved similarly.

We now apply this criterion to our example:

$$G = SL(2, \mathbf{C}) \quad \text{and} \quad V_N = \{ \sum_{i+j=N} a_{ij} X_0^i X_1^j \}.$$

Let's consider

$$\lambda(t) = \begin{pmatrix} t^{-r} & 0 \\ 0 & t^r \end{pmatrix}, \quad r > 0$$

and ask when

$$\lim_{t \to 0} P^{\lambda(t)} = 0.$$

If $P(X_0, X_1) = \sum a_{ij} X_0^i X_1^j$, $P^{\lambda(t)}(X_0, X_1) = \sum a_{ij} X_0^i X_1^j t^{r(i-j)}$. So $\lim_{t\to 0} P^{\lambda(t)} = 0$ means that $a_{ij} = 0$ if $j \geq i$. That is, P has a factor of X_0^k with $k > \frac{N}{2}$. For a general $1-P-S$, we see that P is semistable if and only if P has no linear factors of degree $> \frac{N}{2}$.

LECTURE 3
The Moduli of Stable Bundles

Let $\mathcal{C}$ be a certain class of algebraic objects. For instance, $\mathcal{C}$ could be all smooth projective curves of genus g. Another choice for $\mathcal{C}$ might be the set of all bundles of rank r with fixed numerical characteristics on a fixed projective variety X. To try to form a moduli space, we would like to put the structure of an algebraic variety on $\mathcal{C}$. To use GIT to solve this problem, we select a $C \in \mathcal{C}$. Then we look for a vector space V_C canonically attached to C. For instance, in the case of vector bundles, we can canonically attach to each bundle E on X the vector space $H^0(X,E)$. Generally, we also pick a basis $s_1, ..., s_n$ of V_C and think of some way to attach a point $\Psi_{C,\{s_1,...,s_n\}}$ in $\mathbf{P}^N$ to C and the basis choice so that $SL(n,C)$ operates on $\mathbf{P}^N$ compatibly with change of basis. We then investigate the stability of $\Psi_{C,\{s_1,...,s_n\}}$. If we can show $\Psi_{C,\{s_1,...,s_n\}}$ is stable, then the image of $\Psi_{C,\{s_1,...,s_n\}}$ in $\mathbf{P}^N_s/G$ depends only on C and in good cases determines C.

Example. Let X be a hyperelliptic curve. There is a unique line bundle L on X of degree two with $\dim H^0(X,L) = 2$. A choice of a basis $\{s_0, s_1\}$ of $H^0(X,L)$ provides a map $\phi : X \to \mathbf{P}^1$ by $\phi(x) = (s_0(x), s_1(x))$. ϕ is generically two-to-one, ramified at a set of $2g+2$ distinct points. These points determine a point $\Psi_{X,\{s_0,s_1\}}$ in $\mathbf{P}(V_{2g+2})_s$ by the correspondence between divisors on $\mathbf{P}^1$ and homogeneous polynomials on X_0 and X_1. $\Psi_{X,\{s_0,s_1\}}$ has no multiple factors, so it is stable by the numerical criterion. So for $N \gg 0$,

$$\psi(X) = \Phi_N(\Psi_{X,\{s_0,s_1\}})$$

depends only on X and

$$\psi(X) = \psi(X') \quad \text{iff} \quad X \cong X'.$$

One can show that the set

$$\{\psi(X) | X \text{ hyperelliptic of genus } g\}$$

is an algebraic variety: the moduli space for hyperelliptic curves of genus g.

There are several approaches to using GIT to construct a moduli space of stable bundles on a projective variety X. We will follow the usual convention of algebraic geometry that a vector bundle of rank r is a locally free sheaf E of $\mathcal{O}_X$ modules of rank r. This is equivalent to the usual definition of analytic vector bundle V with analytic transition functions modulo analytic isomorphisms. The correspondence is that E is the sheaf of sections of V. X is embedded in a projective space, so there is a very ample line bundle $\mathcal{O}_X(1) = \mathcal{O}(H)$. If F is a coherent sheaf on X, then there is a polynomial

$$P_F(n) = \sum (-1)^i h^i(X, F(n))$$

called the Hilbert polynomial. For instance, if F is locally free, the Riemann-Roch Theorem tells us how to compute P_F in terms of the Chern classes of F, $\mathcal{O}(1)$ and the tangent bundle of X. If $n \gg 0$ (depending on F) then $P_F(n) = h^0(X, F(n))$. We have the following definition.

Definition. *If E is a torsion free sheaf on X, define*

$$p_E(n) = \frac{P_E(n)}{rk(E)}.$$

Definition. *E is H-semistable if whenever $F \subset E$ is a coherent subsheaf,*

$$p_F(n) \leq p_E(n), \quad n \gg 0.$$

E is stable if the inequality is strict.

We will see later how these definitions arise naturally from GIT. If X is a curve and E is a bundle, then E is H-semistable if and only if whenever $F \subset E$, we have

$$\frac{\deg(F)}{rk(F)} \leq \frac{\deg(E)}{rk(E)}.$$

This follows from $P_E(n) = ndr + \deg(E) + r(1-g)$, where $d = \deg(\mathcal{O}(1))$ and $r = rk(E)$.

Given a polynomial P, how can we construct moduli for the H-stable bundles with Hilbert polynomial P? By this, we mean that if M_P is the set of H-stable bundles with Hilbert polynomial P, we should construct a map $\phi : M_P \to \mathbf{P}^N$ so that if a variety S parameterizes a family of bundles in M_P, the induced map from S to $\mathbf{P}^N$ is an algebraic morphism.

Let $P_k(n) = P(n+k)$. Then there is a bijection between the set of sheaves with Hilbert polynomial P and the set of those with Hilbert polynomial P_k given by

$$E \to E(k).$$

As far as abstract moduli theory is concerned, this correspondence is an equivalence. Now it is a fact that we can choose k so that if E is any H-semistable bundle with polynomial P_k, then $h^i(E) = 0$ for all $i > 0$ and E is generated by global sections. So replacing P by P_k, we can assume that if E is H-semistable, then $h^i(E) = 0$ for all $i > 0$ and E is generated by global sections. We call the process of replacing the set $\{E | P_E = P\}$ by $\{F | P_F = P_k\}$ as shifting the family by k. So we can rephrase by saying that we can assume that the family of H-semistable bundles can be assumed to be globally generated by shifting by k.

Following our philosophy, we start with a bundle E of rank r so that E is generated by global sections and $h^i(E) = 0$ for $i > 0$. Then $H^0(E)$ is a vector space canonically attached to E, of dimension $n = P_E(0)$. Pick a basis $B = \{s_1, ..., s_n\}$ of $H^0(E)$. We want to assign a point $\Psi(E, B) \in \mathbf{P}^N$ which determines E and so that $G = SL(n, \mathbf{C})$ acts on $\mathbf{P}^N$ compatibly with changing the basis B. All the solutions to this problem involve Grassmannians.

Let $G(n,q)$ be the Grassmannian of $n-q$ dimensional subspaces of $\mathbf{C}^n$. $G(n,q)$ is naturally a projective variety via the Plücker embedding as follows: A point P of $G(n,q)$ is equivalent to a quotient V of $\mathbf{C}^n$ of dimension q:

$$\mathbf{C}^n \xrightarrow{\phi} V.$$

Taking the qth exterior power, we have:

$$\wedge^q \mathbf{C}^n \xrightarrow{\wedge^q \phi} \wedge^q V \cong \mathbf{C}.$$

The Plücker coordinates of P are the $\binom{n}{q}$ numbers

$$\wedge^q \phi(e_{i_1} \wedge e_{i_2} \wedge ... \wedge e_{i_q}), \quad i_1 < i_2 < ... < i_q.$$

They are well defined up to a constant independent of $(i_1, i_2, ..., i_q)$, since they only depend on the isomorphism of $\wedge^q V$ with $\mathbf{C}$. If we want to consider subspaces of codim q in a vector space V, we write $\mathrm{Grass}(V, q)$.

The original solution to the moduli of bundles on a smooth projective curve X was obtained by Mumford and Seshadri [**Se**]. We very briefly outline their approach. If $P \in X$, we let E_P denote the fiber of E over P. E_P is a vector space of dimension $rk(E)$. Our basis $B = \{s_1, ..., s_n\}$ thus gives n vectors $s_1(P), ..., s_n(P)$ which span E_P. So we get

$$\mathbf{C}^n \to E_P \to 0,$$

and hence a point $\phi(P) \in G(n,r)$. Mumford and Seshadri fix points $P_1, ..., P_k \in X$ and define

$$\Psi_k(E, B) \to \Pi_k G(n,r) \to \mathbf{P}^L,$$

by $\Psi_k(E, B) = (\phi(P_1), ..., \phi(P_k))$. Now each $G(n,r)$ can be embedded in $\mathbf{P}^{\binom{n}{r}-1}$ and $\Pi_k \mathbf{P}^{\binom{n}{r}-1}$ can be embedded in $\mathbf{P}^L$ by the Segre embedding. We denote the image of $\Psi_k(E, B) \in \mathbf{P}^L$ by $\Psi_k(E, B)$ again. $SL(n, \mathbf{C})$ operates on $\mathbf{P}^L$, so we can discuss the stability of $\Psi_k(E, B)$. The result of Mumford and Seshadri is that if $\deg(E)$ is sufficiently large, $\Psi_k(E, B)$ is GIT semistable if E is semistable as a bundle. We refer to [**N**] for these ideas. However, it is difficult to generalize this approach to higher dimensional varieties. The main reason is that to obtain a compact moduli space, one must include semistable torsion free sheaves and the above approach fails if E is not locally free at one of the P_i.

The second approach to the moduli of bundles works for any smooth projective variety X. Instead of analyzing all bundles of Hilbert polynomial P, we fix a line bundle L and consider only those bundles E with Hilbert polynomial P and $\det(E) = L$, where $\det(E)$ is the highest exterior power of E. Since L is fixed, we can identify $H^0(L)$ with $\mathbf{C}^K$, where $K = h^0(L)$. Giving a basis B of $H^0(E)$ is the same as an isomorphism

$$\phi : \mathbf{C}^n \to H^0(E).$$

There is a natural map

$$\wedge^r H^0(X, E) \to H^0(X, \wedge^r E)$$

given by taking sections $s_1, ..., s_r$ of E and wedging them together to get a section of $\wedge^r E$. Choose an isomorphism of $\wedge^r E$ with L, which is determined up to an element of $\mathbf{C}^*$. We define

$$T_{E,\phi} : \wedge^r \mathbf{C}^n \to \mathbf{C}^K$$

to be the composition

$$\wedge^r \mathbf{C}^n \xrightarrow{\wedge^r \phi} \wedge^r H^0(E) \longrightarrow H^0(\wedge^r E) \longrightarrow H^0(L) = \mathbf{C}^K.$$

Thus we get

$$T_{E,\phi} \in \mathbf{P}(\mathrm{Hom}(\wedge^r \mathbf{C}^n, \mathbf{C}^K))$$

which is independent of choices of isomorphisms.

When E is torsion free, we can still define $\det(E)$ as a line bundle. For E is locally free in codimension one and there is a unique line bundle $\det(E)$ coincides with $\wedge^r E$ where E is locally free. So $T_{E,\phi}$ can be defined even if E is torsion free. Note that $T_{E,\phi}$ determines E if E is a bundle. For let $s_1, ..., s_n$ be a basis of $H^0(E)$. Then from $T_{E,\phi}$, we can determine the sections $s_{i_1} \wedge ... \wedge s_{i_r} \in H^0(\det(E))$. However, $s_{i_1}(P) \wedge ... \wedge s_{i_r}(P)$ are just the Plücker coordinates of $\mathbf{C}^n \to E_P \to 0$. Thus we can recover the sheaf $K = \mathrm{Ker}(\mathcal{O}^n \to E)$, since we know $K_P \subseteq \mathcal{O}_P^n$ for each P.

One of the main results of [**G**], [**M1**] and [**M2**] is that if P is a fixed polynomial and L a fixed line bundle, then there is a k so that if E is any H-semistable torsion free sheaf with $P_E = P$ and $\det(E) = L$, then $T_{E,\phi}$ is semistable for any choice of an isomorphism $\phi : H^0(E(k)) \cong \mathbf{C}^{P(k)}$.

The proof of this result is long and rather technical. To give a flavor of the proof, we give the proof for rank two bundles on a smooth projective curve X. So we let E be a rank two stable bundle on a smooth projective curve X, $\det(E) = L$. Suppose $\deg(L) \gg g = g(X)$. Then the stability of E will imply that E is globally generated and $H^1(E) = 0$. Fix a basis $\{s_1, ..., s_n\}$ of $H^0(E)$, which corresponds to an isomorphism

$$\phi : \mathbf{C}^n \longrightarrow H^0(E).$$

Then to the pair (E, ϕ) is attached a point

$$T_{E,\phi} \in \mathbf{P}(\mathrm{Hom}(\wedge^2 \mathbf{C}^n, \mathbf{C}^K))$$

as above. We want to show that $T_{E,\phi}$ is stable in the sense of GIT. Suppose it is not. Then there is a $1-P-S$ λ so that $T_{E,\phi}$ is not λ stable. Let $v_1, ..., v_n$ be a basis of $H^0(E) = \mathbf{C}^n$ so that $v_i^{\lambda(\alpha)} = \alpha^{r_i} v_i$. Then $v_i \wedge v_j = 0$ in $H^0(\wedge^2 E)$ if $r_i + r_j < 0$. We can assume $r_1 \leq r_2 \leq ... \leq r_n$. Let M be the subbundle of E so that $v_1 \in H^0(M)$. Note that $v_1 \wedge v_j = 0$ if $r_1 + r_j < 0$. Thus $v_j \in H^0(M)$ if $r_1 + r_j < 0$.

Let $N = \#\{j | r_j < |r_1|\}$. Then $h^0(M) \geq N$. If $h^1(M) \neq 0$, then $h^0(M) \leq g$; while if $h^1(M) = 0$, then $h^0(M) = \deg(M) + 1 - g < \frac{1}{2}(\deg(E) + 2(1-g)) = \frac{n}{2}$, since E is stable. Thus $N < \frac{n}{2}$ in both cases. For more than $\frac{n}{2}$ of the j, we have $r_j > |r_1|$. This contradicts $\sum r_i = 0$.

The third approach to moduli of bundles is due to Simpson [**Si**]. Let E be a coherent sheaf with Hilbert polynomial $P_E = P$ and suppose $s_1, ..., s_n \in H^0(E)$ are so that the map

$$\mathcal{O}^n \xrightarrow{\phi} E \to 0$$

is surjective as a map of sheaves. Grothendieck showed that we can find an l depending only on P so that

$$H^0(\mathcal{O}(l)^n) \to H^0(E(l)) \to 0.$$

For a fixed sheaf, this just follows from the long exact sequence associated to

$$0 \to K(l) \to \mathcal{O}(l)^n \to E(l) \to 0$$

and the fact that $H^1(K(l)) = 0$ for $l \gg 0$. Let $V_l = H^0(\mathcal{O}(l))$. Then

$$H^0(\mathcal{O}(l)^n) = \mathbf{C}^n \otimes V_l.$$

So it defines a point

$$\Phi_l(E, \phi) \in \text{Grass}(\mathbf{C}^n \otimes V_l, P_E(l)),$$

which we call the lth Hilbert point of (E, ϕ). One interesting result of Grothendieck is that $\{\Phi_l(E, \phi)\}$ is projective, i.e., it has a natural structure as a projective scheme. Further, $\Phi_l(E, \phi) = \Phi_l(E', \phi')$ if and only if ϕ and ϕ' have the same kernel. Indeed, if $K = \text{Ker}(\phi : \mathcal{O}^n \to E)$, then Φ_l determines $H^0(K(l)) \subseteq H^0(\mathcal{O}(l)^n)$. But for $l \gg 0$, $K(l)$ is generated by its global sections. So $K(l)$ can be reconstructed as subsheaf of $\mathcal{O}(l)^n$ generated by the sections of $H^0(K(l))$. So $K(l)$ is determined by $H^0(K(l))$, which in turn is determined by $\Phi_l(E, \phi)$. Now $SL(n, \mathbf{C})$ operates on $\text{Grass}(\mathbf{C}^n \otimes V_l, P_E(l))$ by

$$\Phi_l(E, \phi)^A = \Phi_l(E, A \circ \phi \circ A^{-1}),$$

where we think of $A^{-1} : \mathcal{O}^n \to \mathcal{O}^n$. If $B = \{s_1, ..., s_n\}$ is a basis of $H^0(E)$ and E is generated by global sections, we denote $\Phi_l(E, \phi)$ by $\Phi_l(E, B)$ where ϕ is the map $\mathcal{O}^n \to E \to 0$ induced by the s_i's. For this type of sheaf E, the orbit of $\Phi_l(E, B)$ only depends on E.

Theorem (Simpson). *After replacing P by P_k, $k \gg 0$ and choosing $l \gg 0$, $\Phi_l(E, B)$ is semistable if E is H-semistable.*

The first step in establishing Simpson's theorem is to determine when a point R of $\text{Grass}(\mathbf{C}^n \otimes V_l, P_E(l))$ is stable or semistable. For notational convenience, we simply write V for V_l and P for $P_E(l)$. Let λ be a $1 - P - S$ of $SL(n, \mathbf{C})$ and $\{w_i\}$

a basis of $\mathbf{C}^n$ so that $w_i^{\lambda(\alpha)} = \alpha^{r_i} w_i$ with $r_1 \leq \ldots \leq r_n$. We say R is λ stable if $\lim_{\alpha \to 0} R^{\lambda(\alpha)} = \infty$. R is just a map

$$\mathbf{C}^n \otimes V \xrightarrow{\phi} W \longrightarrow 0,$$

where $\dim W = P$. Let $\{v_i\}$ be a basis for V. The homogeneous coordinates of R are just

$$(\wedge^P \phi)((w_{i_1} \otimes v_{j_1}) \wedge \ldots \wedge (w_{i_P} \otimes v_{j_P})),$$

and

$$\begin{aligned}&\left((\wedge^P \phi)((w_{i_1} \otimes v_{j_1}) \wedge \ldots \wedge (w_{i_P} \otimes v_{j_P}))\right)^{\lambda(\alpha)} \\ =&\alpha^{r_{i_1} + \ldots + r_{i_P}} (\wedge^P \phi)((w_{i_1} \otimes v_{j_1}) \wedge \ldots \wedge (w_{i_P} \otimes v_{j_P})).\end{aligned}$$

Thus $\lim_{\alpha \to 0} R^{\lambda(\alpha)} = \infty$ if and only if there are $i_1, \ldots, i_p, j_1, \ldots, j_p$ so that

$$(\wedge^P \phi)((w_{i_1} \otimes v_{j_1}) \wedge \ldots \wedge (w_{i_P} \otimes v_{j_P})) \neq 0 \quad \text{and} \quad r_{i_1} + \ldots + r_{i_p} < 0.$$

Now

$$(\wedge^P \phi)((w_{i_1} \otimes v_{j_1}) \wedge \ldots \wedge (w_{i_P} \otimes v_{j_P})) \neq 0$$

just says that $\phi(w_{i_1} \otimes v_{i_1}), \ldots, \phi(w_{i_p} \otimes v_{i_p})$ are a basis for W. So to show that R is stable, all we have to do is to find $w_{i_1}, \ldots, w_{i_p}, v_{i_1}, \ldots, v_{i_p}$ with $\sum r_{i_j} < 0$ and $\phi(w_{i_1} \otimes v_{i_1}), \ldots, \phi(w_{i_p} \otimes v_{i_p})$ a basis for E. How should we choose such a basis? One way is first to use w_1 as much as possible, as w_1 has the least r, then to use w_2 as much as possible, etc. Let E_i be the span of $w_1, \ldots, w_i$ in $\mathbf{C}^n$ and let H_i be the image of $E_i \otimes V$ in W and let $h_i = \dim H_i$.

Proposition (Mumford). *R is λ stable if and only if for all $k < n$,*

$$\frac{k}{h_k} < \frac{n}{h_n}.$$

Proof. Suppose

$$\frac{k}{h_k} < \frac{n}{h_n}.$$

Pick

$$w_1 \otimes v_{i_1}, \ldots, w_1 \otimes v_{i_{h_1}}, w_2 \otimes v_{i_{h_1+1}}, \ldots, w_2 \otimes v_{i_{h_2}}, \ldots$$

so that

$$\phi(w_1 \otimes v_{i_1}), \ldots, \phi(w_1 \otimes v_{i_{h_1}}), \phi(w_2 \otimes v_{i_{h_1+1}}), \ldots, \phi(w_2 \otimes v_{i_{h_2}}), \ldots$$

form a basis of W. So $r_{i_1} = \ldots = r_{i_{h_1}} = r_1$, $r_{i_{h_1+1}} = \ldots = r_{i_{h_2}} = r_2, \ldots$ and

$$\sum r_{i_j} = r_1 h_1 + r_2(h_2 - h_1) + \ldots + r_n(h_n - h_{n-1}).$$

Let $L(q_1, \ldots, q_n) = q_1 h_1 + q_2(h_2 - h_1) + \ldots + q_n(h_n - h_{n-1})$. If R is not stable, then $L(r_1, \ldots, r_n) \geq 0$. Let K be the compact set

$$K = \{(q_1, \ldots, q_n) | q_i \in \mathbf{R}, \sum q_i = 0, \text{ and } q_1 = -1, q_1 \leq \ldots \leq q_n\}.$$

Then L is nonnegative at some point of K, namely

$$(-\frac{r_1}{r_1}, -\frac{r_2}{r_1}, ..., -\frac{r_n}{r_1}).$$

K is defined by linear inequalities, so by linear programming, L takes its maximum value at a vertex of K. At this point,

$$-1 = q_1 = ... = q_k < q_{k+1} = ... = q_n = \frac{k}{n-k},$$

and $L(q_1, ..., q_n) =$

$$-h_1 - (h_2 - h_1) - ... - (h_k - h_{k-1}) + \frac{k}{n-k}(h_{k+1} - h_k) + ... + \frac{k}{n-k}(h_n - h_{n-1}) \geq 0.$$

So

$$h_k \leq \frac{k}{n-k}(h_n - h_k), \quad \text{i.e.,} \quad \frac{k}{h_k} \geq \frac{n}{h_n}.$$

This contradicts our assumption. So R is λ stable. Conversely, if

$$\frac{k}{h_k} \geq \frac{n}{h_n},$$

we define $r_1 = ... = r_k = k-n$, $r_{k+1} = ... = r_n = k$. We see that $\lim_{\alpha \to 0} R^{\lambda(\alpha)} \neq \infty$, so R is not λ stable. If we replace " $<$ " by " $\leq$ " in the proposition, we have the statement for semistability.

The second step in establishing Simpson's theorem is to investigate the following situation. Let E be H-semistable and $W \subseteq H^0(E)$ is a subspace of dimension p. We want to estimate the dimension $d_{W,l}$ of $W \otimes H^0(\mathcal{O}(l))$ in $H^0(E(l))$. Let F be the subsheaf of E generated by sections in W. We have the following exact sequence:

$$0 \to K(l) \to W \otimes \mathcal{O}(l) \to F(l) \to 0.$$

For $l \gg 0$, the map

$$W \otimes V_l = H^0(W \otimes \mathcal{O}(l)) \longrightarrow H^0(F(l))$$

is surjective. So $d_{W,l} = h^0(F(l))$.

Once again, we work this out for the case when X is a smooth projective curve and E a semistable bundle of rank two. Further, we assume $\deg(E) \gg g = g(X)$, so $h^0(E) = \deg(E) + 2(1-g)$. The numerical criterion (i.e., the above proposition) shows that

$$H^0(E) \otimes H^0(\mathcal{O}(l)) \xrightarrow{\phi} H^0(E(l)) \longrightarrow 0$$

is semistable if and only if whenever $W \subseteq H^0(E)$ is a subspace of dimension p, we have

$$\frac{\dim \phi(W \otimes H^0(\mathcal{O}(l)))}{p} \geq \frac{h^0(E(l))}{h^0(E)},$$

i.e.,

$$\frac{\dim \phi\big(W \otimes H^0(\mathcal{O}(l))\big)}{p} \geq \frac{2la + \deg(E) + 2(1-g)}{\deg(E) + 2(1-g)},$$

where $a = \deg(\mathcal{O}(1))$. Let's assume the sheaf F generated by W is a line bundle. Then $\dim \phi\big(W \otimes H^0(\mathcal{O}(l))\big) = al + \deg(F) + 1 - g$.
Case 1: $h^1(F) = 0$. Then $p \leq h^0(F) = \deg(F) + 1 - g$. So

$$\frac{\dim \phi\big(W \otimes H^0(\mathcal{O}(l))\big)}{p} = \frac{al + \deg(F) + 1 - g}{p} \geq \frac{al + \deg(F) + 1 - g}{\deg(F) + 1 - g}.$$

Since E is semistable, $\deg(F) \leq \frac{1}{2}\deg(E)$. So we can arrange

$$\frac{al + \deg(F) + 1 - g}{\deg(F) + 1 - g} \geq \frac{2la + \deg(E) + 2(1-g)}{\deg(E) + 2(1-g)}$$

by choosing $l \gg 0$.
Case 2: $H^1(F) \neq 0$. In this case, $\deg(F) < 2g - 1$ and $\deg(F) + h^1(F) \leq 2g - 1$. So by choosing $\deg(E) \gg 0$, we may assume $\deg(F) < \frac{1}{2}\deg(E) - h^1(F) - 1$. Now $p \leq h^0(F) = h^1(F) + \deg(F) + (1-g)$. So $p \leq \frac{1}{2}\deg(E) + (1-g) - 1 = \frac{1}{2}(\deg(E) + 2(1-g) - 2)$, implying that

$$\frac{\dim \phi\big(W \otimes H^0(\mathcal{O}(l))\big)}{p} = \frac{al + \deg(F) + 1 - g}{p} \geq \frac{2la + \deg(E) + 2(1-g)}{\deg(E) + 2(1-g)}$$

for $l \gg 0$. Therefore, (E, ϕ) is GIT semistable in both cases. We leave it to the reader to prove that (E, ϕ) is GIT semistable if F is a vector bundle.

References

[G] D. Gieseker, *On the moduli of vector bundles on an algebraic surface*, Annals of Math. **106** (1977), 45.

[L] S. Lang, *Algebra*, Third Edition, Addison-Wesley, Reading, MA., 1993.

[M1] M. Maruyama, *Moduli of Stable Sheaves* I, J. Math. Kyoto **17** (1977), 91.

[M2] M. Maruyama, *Moduli of Stable Sheaves* II, J. Math. Kyoto **18** (1978), 557.

[MFK] D. Mumford, J. Fogarty, and F. Kirwan, *Geometric Invariant Theory*, Third edition, Springer, Berlin, 1994.

[N] P. E. Newstead, *Lectures on introduction to moduli problems and orbit spaces*, Springer, Berlin, 1978.

[Se] C. S. Seshadri, *Space of unitary vector bundles on a compact Riemann surface*, Annals of Math. **85** (1972), 303.

[Si] C. Simpson, *Higgs bundles and local systems*, Publ. Math. IHES **74**, 5.

Anti-Self-Dual Connections and Stable Vector Bundles

Jun Li

IAS/Park City Mathematics Series
Volume 4, 1998

Anti-Self-Dual Connections and Stable Vector Bundles

J. Li

Introduction

In this lecture series, I will discuss briefly the celebrated theorem of Donaldson and Kobayashi showing that the space of irreducible Anti-Self-Dual connections (modulo equivalence) is isomorphic to the moduli of stable bundles. According to gauge theory, one can define invariants of smooth structures of four-manifolds by investigating global structures of the moduli of Anti-Self-Dual connections. This theorem transforms the problem of studying moduli of Anti-Self-Dual connections (over algebraic surfaces) into problem of studying moduli of stable vector bundles, in effect allowing one to use algebraic method to study these invariants.

[1]Department of Mathematics, Stanford University, Stanford, CA 94305
E-mail address: jli@gauss.stanford.edu

LECTURE 1
Hermitian Bundles, Hermitian Connections and Their Curvatures

Recall that a principal SU(2)-bundle P over a manifold X is a manifold with a smooth free SU(2) action and orbit space $P/\mathrm{SU}(2) = X$. On P one can introduce the notion of SU(2)-connections, which are SU(2)-invariant splitting of the tangent bundle TP into horizontal and vertical spaces. (For details please see J. Morgan's lecture notes in this series.) For the purpose of our investigation, we will work with the associated vector bundles of principal bundles, which is convenient when the underlying manifold is a Kähler surface.

We fix a principal SU(2)-bundle P over X. Let V be a complex 2-dimensional linear space with a fixed Hermitian inner product $< \cdot, \cdot >_V$ and a volume form $\mathbf{1} \in \wedge^2 V$. Let

$$\rho : \mathrm{SU}(2) \longrightarrow \mathrm{GL}(V)$$

be a representation preserving the Hermitian product and the volume form. Then we obtain a vector bundle over X defined by taking the product

$$E = P \times_{SU(2)} V = P \times V/_{\sim},$$

where $(z, v) \sim (z', v')$ if there is a $g \in \mathrm{SU}(2)$ such that $(z', v') = (zg, g^{-1}v)$. Because the SU(2) representation preserves the volume form $\mathbf{1} \in \wedge^2 V$,

$$\wedge^2 E = P \times_{SU(2)} \wedge^2 V = \mathbf{C}_X$$

is the trivial line bundle. Hence the first Chern class of E, which is the first Chern class of $\wedge^2 E$, is trivial. We denote $\pi : E \to X$ the projection and for any $x \in X$ we denote $\pi^{-1}(x)$ by E_x. E is canonically a Hermitian vector bundle in the following sense.

Definition 1. A Hermitian vector bundle E is a vector bundle coupled with a Hermitian metric h. A Hermitian metric h is a collection of Hermitian products

$$h_x : E_x \times E_x \longrightarrow \mathbb{C}, \quad x \in X$$

such that $h_x(\cdot, \cdot)$ is smooth in $x \in X$.

Indeed, the Hermitian product on V induces a Hermitian metric on $E = P \times_{SU(2)} V$ since the representation of SU(2) in V preserves the given Hermitian product $< \cdot, \cdot >_V$.

Exercise 1. Let (E, h) be a Hermitian vector bundle. Show that (E, h) is the associated vector bundle of a principal U(2)-bundle. (Hint: Consider the orthonormal frame bundle of (E, h).)

Exercise 2. Let (E, h) be as in the Exercise 1. Assume further that we have an isomorphism $\wedge^2 E \cong \mathbf{C}_X$ and under this isomorphism $\wedge^2 h = 1$. Then (E, h) is the associated vector bundle of an SU(2)-principal bundle.

There is a local description of vector bundles as follows: Let $\pi : E \longrightarrow X$ be any vector bundle. Then there is a local covering $\{U_\alpha\}$ of X so that there are trivializations of the restriction of E to U_α

$$E|_{U_\alpha} = \pi^{-1}(U_\alpha) \cong U_\alpha \times \mathbf{C}^2 \tag{1}$$

and transition functions

$$\phi_{\alpha\beta} : U_\alpha \cap U_\beta \longrightarrow GL(\mathbb{C}^2), \tag{2}$$

satisfying the obvious compatibility conditions. Since (E, h) is a Hermitian vector bundle, h induces Hermitian metric h_α on $E|_{U_\alpha} = U_\alpha \times \mathbf{C}^2$, which are compatible in the sense that over $U_\alpha \cap U_\beta$

$$h_\beta(\cdot,\cdot) = h_\alpha(\phi_{\alpha\beta}(\cdot), \phi_{\alpha\beta}(\cdot)).$$

Of course, we can choose (1) so that h_α is the standard Hermitian product on $U_\alpha \times \mathbf{C}^2$. To avoid any confusion on notations, in the following we will express vectors in $\mathbf{C}^2$ as column vectors, $\phi_{\alpha\beta}$ as 2×2-matrices acting on v from the left and $h_\alpha(\xi, \eta) = \xi^t h_\alpha \bar{\eta}$.

From now on, we assume X is a smooth complex surface. We will introduce notion of Hermitian connections and curvatures of (E, h). Recall that a connection D on E is a differential operator

$$D : \Omega^0(E) \longrightarrow \Omega^1(E)$$

satisfying the Leibnitz rule: For any smooth function f and section v of E,

$$D(fv) = df \cdot v + f \cdot D(v).$$

When we work with Hermitian bundle (E, h), we certainly like D to preserve the Hermitian structure h in the following sense.

Definition 2. A connection D on (E, h) is a Hermitian connection if

$$d\,h(\xi, \eta) = h(D\xi, \eta) + h(\xi, D\eta)$$

holds for any sections $\xi, \eta \in \Omega^0(E)$.

Under the local trivialization $E|_{U_\alpha} \cong U_\alpha \times \mathbf{C}^2$, any connection D can be expressed as

$$D_\alpha = d + A_\alpha,$$

where $A_\alpha \in \Omega^1(U_\alpha, E^\vee \otimes E)$ is a 1-form with value in the vector bundle $E^\vee \otimes E$.

Exercise 3. Show that if $D_\alpha = d + A_\alpha$ and $D_\beta = d + A_\beta$ are local expressions over U_α and U_β of a connection D respectively, then over $U_\alpha \cap U_\beta$ we have

$$\phi_{\alpha\beta}^{-1} d\phi_{\alpha\beta} + \phi_{\alpha\beta}^{-1} A_\alpha \phi_{\alpha\beta} = A_\beta,$$

where $\phi_{\alpha\beta}$ is the transition function in (2).

Exercise 4. Show that if D is a Hermitian connection of (E, h), then the local form $D_\alpha = d + A_\alpha$ must satisfy

$$A_\alpha^t h_\alpha + h_\alpha \overline{A}_\alpha = d\, h_\alpha,$$

where h_α is the local form of h.

Similar to the case of principal bundle, the space of Hermitian connections on (E, h) is an infinite dimensional affine space. This can be seen by noticing that if D is a Hermitian connection and $\omega \in \Omega^1(E^\vee \otimes E)$ is skew-symmetric, i.e.

$$h(\omega\, \xi, \eta) + h(\xi, \omega\, \eta) = 0,$$

then $D + \omega$ is a Hermitian connection. In particular, there are no preferred Hermitian connections on (E, h). However, if we bring in the holomorphic structure on E, we get a different picture.

Definition 3. A holomorphic structure on a vector bundle E consists of an open covering $\{U_\alpha\}$ of X and a local trivialization

$$E|_{U_\alpha} \cong U_\alpha \times \mathbf{C}^2$$

such that the transition functions $\phi_{\alpha\beta}$ are all holomorphic.

Let E be a holomorphic vector bundle. We can define $\overline{\partial} v$ for any section v of E as follows: Let $v_\alpha = v|_{U_\alpha}$ be the local form under $E|_{U_\alpha} \cong U_\alpha \times \mathbf{C}^2$. Since all transition functions are holomorphic, $\overline{\partial} v_\alpha = \phi_{\alpha\beta} \overline{\partial} v_\beta$ over $U_\alpha \cap U_\beta$. Therefore, all $\overline{\partial} v_\alpha$'s fit together to form a global $\overline{\partial} v \in \Omega^{0,1}(E)$. In this way, we obtain a $\overline{\partial}$-operator

$$\overline{\partial} : \Omega^0(E) \longrightarrow \Omega^{0,1}(E).$$

Obviously, $\overline{\partial}$ satisfies the Leibnitz rule and $\overline{\partial} \circ \overline{\partial} = 0$. In fact, the converse is also true.

Proposition 4. *Let E be a (smooth) vector bundle. Then any operator*

$$\overline{\partial} : \Omega^0(E) \longrightarrow \Omega^{0,1}(E)$$

satisfying the $\overline{\partial}$-Leibnitz rule

$$\overline{\partial}(fv) = \overline{\partial} f \cdot v + f \overline{\partial}$$

and the integrability condition $\overline{\partial} \circ \overline{\partial} = 0$ is the $\overline{\partial}$-operator of a holomorphic structure on E.

A proof of this Proposition can be found on page 51 of [**DK**]. In the following, we will not distinguish between a holomorphic structure and an integrable $\overline{\partial}$-operator. We understand that a holomorphic vector bundle is a vector bundle with a holomorphic structure.

In the following, we will investigate how to find a canonical Hermitian connection of a holomorphic hermitian vector bundle $(E, h, \overline{\partial})$. Since complex structure on X induces splitting $\Omega^1 = \Omega^{1,0} \oplus \Omega^{0,1}$, any connection D can be written as a direct sum

$$D = D' + D'',$$

where $D' : \Omega^0(E) \to \Omega^{1,0}(E)$ and $D'' : \Omega^0(E) \to \Omega^{0,1}(E)$.

Proposition 5. *Any Hermitian holomorphic vector bundle $(E, h, \overline{\partial})$ admits a unique Hermitian connection D such that $D'' = \overline{\partial}$.*

Proof. Let $E|_{U_\alpha} = U_\alpha \times \mathbf{C}^2$ be a (holomorphic) trivialization of the holomorphic vector bundle E and let h_α be the induced Hermitian structure on $E|_{U_\alpha}$. Assume $D_\alpha = d + A_\alpha$ is the restriction to U_α of the desired connection D, then necessarily A_α is of (1,0)-type, because $D'' = \overline{\partial}$. Further for any section ξ and η of E_α,

$$d(\xi^t h_\alpha \bar{\eta}) = (d\xi + A_\alpha \xi)^t h_\alpha \bar{\eta} + \xi^t h_\alpha \overline{(d\eta + A_\alpha \eta)},$$

since D is Hermitian. After simplification, we get

$$dh_\alpha = A_\alpha^t h_\alpha + h_\alpha \overline{A}_\alpha.$$

Because A_α is of (1,0)-type, it must take the form

$$A_\alpha = h_\alpha^{-1} \partial h_\alpha. \tag{3}$$

This shows that if D exists, then it must be unique. Based on (3), one verifies that these (local) connections $d + A_\alpha$ fit together to form the desired connection D. We call this unique connection *the Hermitian connection* of the holomorphic Hermitian bundle $(E, h, \overline{\partial})$.

As usual, we define the curvature $\Theta(D)$ of any connection D to be

$$\Theta(D) = D \circ D \in \Omega^2(E^\vee \otimes E). \tag{4}$$

Proposition 6. *1. The curvature of the Hermitian connection of a holomorphic Hermitian vector bundle is of (1,1)-type. 2. If (E, h) is a smooth Hermitian vector bundle and if D is a Hermitian connection such that its curvature $\Theta(D)$ is of (1,1)-type, then there is a unique holomorphic structure on E making D the Hermitian connection of this holomorphic Hermitian bundle.*

Proof. First of all since

$$\begin{aligned}\Theta(D) &= D \circ D = (D' + D'') \circ (D' + D'') \\ &= D' \circ D' + (D' \circ D'' + D'' \circ D') + D'' \circ D''\end{aligned}$$

and since $\overline{\partial} \circ \overline{\partial} = D'' \circ D'' = 0$, the (0,2)-part of Θ vanishes identically. On the other hand, because $d(dh(\xi, \eta)) = 0$ we have

$$\begin{aligned}0 &= d(h(D\xi, \eta) + h(\xi, D\eta)) \\ &= (h(D^2\xi, \eta) - h(D\xi, D\eta)) + (h(D\xi, D\eta) + h(\xi, D^2\eta)).\end{aligned}$$

Hence,

$$h(\Theta\xi, \eta) + h(\xi, \Theta\eta) = 0.$$

Therefore, the (2,0)-part of Θ vanishes as well. This proves the first part of the proposition.

Now we prove the second part of the proposition. Assume $\Theta(D) \in \Omega^{1,1}(E^\vee \otimes E)$. Let D'' be the (0,1)-part of the connection. Then D'' satisfies the $\overline{\partial}$-Leibnitz rule and further, $D'' \circ D''$ is the (0,2)-part of the connection Θ, which vanishes by the assumption. By Proposition 4, there is a holomorphic structure $\overline{\partial}$ on E so that $\overline{\partial} = D''$. D will be the Hermitian connection of the holomorphic Hermitian vector bundle $(E, h, \overline{\partial})$.

LECTURE 2
Hermitian-Einstein Connections and Stable Vector Bundles

In this lecture, we will introduce the notion of H-E connections and show that bundles with irreducible H-E connections are slope-stable. We first analyze the notion of Anti-Self-Dual connections in the context of Kähler geometry. Let X be a smooth Kähler surface with Kähler form ω and the associated Riemannian metric g_ω. Let P be an SU(2)-principal bundle with SU(2)-connection $\tilde{D}$. As was mentioned in the first lecture, P associates to a unique smooth rank 2 Hermitian vector bundle (E, h) with $\wedge^2 E \cong \mathbf{C}_X$. The SU(2) connection $\tilde{D}$ will induce canonically a Hermitian connection D whose curvature $\Theta(D)$ is identical to the curvature $\Theta(\tilde{D})$ under the inclusion of vector bundles

$$ad\, P \subset E^\vee \otimes E.$$

Now we assume $\tilde{D}$ is an ASD-connection. Namely, $\Theta(\tilde{D}) \in \Gamma\big(\Omega^2_-(adP)\big)$, where Ω^2_- is the bundle of Anti-Self-Dual 2-forms with respect to g_ω. (For the details of Anti-Self-Dual connection and the related topics, please see the lecture notes of J. Morgan in this series.) Then we must have

$$\Theta(D) \in \Gamma\big(\Omega^2_-(E^\vee \otimes E)\big). \tag{5}$$

For any $x \in X$, we choose a local (holomorphic) coordinate (z_1, z_2) of X at x so that the restriction of the Kähler form ω at x is

$$\omega_x = \frac{i}{2}(dz_1 \wedge d\bar{z}_1 + dz_2 \wedge d\bar{z}_2). \tag{6}$$

Lemma 7. *Assume the Kähler form ω has the form (6) at x. Then the Self-Dual forms at x are linear combinations of*

$$dz_1 \wedge dz_2,\quad d\bar{z}_1 \wedge d\bar{z}_2 \quad \textit{and} \quad dz_1 \wedge d\bar{z}_1 + dz_2 \wedge d\bar{z}_2$$

and the Anti-Self-Dual forms at x lie in the orthogonal complement of ω_x in $\Omega^{1,1}_x$.

Exercise 5. Verify lemma 7 directly.

A direct consequence of this lemma is the following useful criterion.

Proposition 8. *Let $\tilde{D}$ be an SU(2)-connection on P and D be the associated Hermitian connection on the associated Hermitian bundle (E, h). Then $\tilde{D}$ is ASD if and only if*

1. *$\Theta(D)$ is of (1,1)-type and*
2. *$\Theta(D) \wedge \omega = 0$.*

Definition 9. Let F be a holomorphic vector bundle on the Kähler surface (X, ω). If h is a Hermitian metric on F whose Hermitian connection D satisfies

$$\Theta(D) \wedge \omega = c\omega^2 \cdot I_F, \tag{7}$$

where c is a constant and I_F is the identity endomorphism of F, then we call D a Hermitian-Einstein (H-E in short) connection of F and call h a Hermitian-Einstein metric.

Note that if $c_1(F) = 0$, which is the case when F is the associated Hermitian bundle of an SU(2)-bundle, then necessarily $c = 0$.

Proposition 10. *Let $\tilde{D}$ be an SU(2) connection on P and D the induced connection on (E, h), the associated hermitian bundle of P. Then $\tilde{D}$ is ASD if and only if there is a holomorphic structure $\overline{\partial}_{\tilde{D}}$ on E such that D is the hermitian connection of $(E, \overline{\partial}_{\tilde{D}})$ and D (and h) is H-E. The correspondence $\tilde{D} \mapsto \overline{\partial}_{\tilde{D}}$ is one-to-one.*

Proof. Let D be the induced connection on the associated Hermitian vector bundle (E, h). D is Hermitian. Since $\tilde{D}$ is Anti-Self-Dual, $\Theta(D)$ is of (1,1)-type. Hence by Proposition 5 there is a holomorphic structure $\overline{\partial}_{\tilde{D}}$ on E so that $D'' = \overline{\partial}_{\tilde{D}}$. Therefore, D is the hermitian connection of $(E, h, \overline{\partial}_{\tilde{D}})$. By Proposition 8 the Hermitian connection D is H-E. The last statement of the Proposition follows from the uniqueness of the Hermitian connection of a holomorphic Hermitian vector bundle.

Because E is the associated vector bundle of P, it is straightforward to check that if $\tilde{D}_1$ and $\tilde{D}_2$ are two gauge equivalent ASD connections, then there is a smooth automorphism $g: E \to E$, preserving h and the volume form $\mathbf{1} \in \wedge^2 E$, such that

$$g^*(\overline{\partial}_{\tilde{D}_1}) \overset{\text{def.}}{=\!=} g^{-1} \circ \overline{\partial}_{\tilde{D}_1} \circ g = \overline{\partial}_{\tilde{D}_2}.$$

(We call such g SU(2)-gauge transformations of E. When g are merely automorphisms, then g are called GL(2)-gauge transformations.) Conversely, assume $\overline{\partial}$ is a holomorphic structure on (E, h) whose Hermitian connection is irreducible and H-E, then the induced connection $\tilde{D}$ on P is necessarily irreducible and ASD. In this way, we get a one-to-one correspondence between the space (resp. the space of gauge equivalent classes) of irreducible ASD connections on P and the space (resp. the space of gauge equivalent classes) of holomorphic structures on (E, h) whose Hermitian connections are irreducible and H-E.

To establish a bridge between gauge theory and algebraic geometry, we need to reinterpret the condition of having H-E connections in terms of algebraic condition on the holomorphic vector bundle $(E, \overline{\partial})$. As was indicated in Friedman and Gieseker's lecture, the right class of holomorphic structures are those satisfying the so called stability condition.

Definition 11. A rank 2 holomorphic vector bundle E is said to be slope-stable[2] if whenever $\mathcal{L} \subset \mathcal{O}(E)$ is a rank one subsheaf, then

$$\deg \mathcal{L} < \frac{1}{2} \deg \mathcal{O}(E),$$

where the degree of an arbitrary sheaf $\mathcal{F}$ is defined to be $\int_X c_1(\mathcal{F}) \wedge \omega$. Here $\mathcal{O}(E)$ is the sheaf of holomorphic sections of the holomorphic vector bundle E.

Remark. In the definition of (rank 2) slope-stable vector bundles, we can replace condition on subsheaves by the following equivalent condition: For any holomorphic line bundle L on X having non-trivial sections in $L^\vee \otimes E$, then

$$\deg L < \frac{1}{2} \deg E.$$

The above two definitions are equivalent because any subsheaf $\mathcal{L}$ of $\mathcal{O}(E)$ is the image sheaf of a (sheaf) homomorphism

$$\mathcal{O}(L) \longrightarrow \mathcal{O}(E),$$

where L is some line bundle, which is given by a non-trivial section of $L^\vee \otimes E$.

Now we state the celebrated theorem that establishes a correspondence between ASD connections and slope-stable bundles.

Theorem 12. *a) Assume E is a holomorphic bundle that admits an irreducible H-E connection, then E is necessarily slope-stable. b) Assume E is a slope-stable holomorphic bundle, then there is a Hermitian metric h (unique up to scalar) on E such that whose connection is irreducible and H-E.*

The first part of the theorem was discovered by Kobayashi. Part 2 was established by Donaldson for algebraic surfaces and was generalized to arbitrary (high dimensional) Kähler manifolds by Uhlenbeck and Yau.

Before going into the details of the proof, let us first explore its implication in gauge theory. We first look at the set

$$\left\{ \begin{matrix} \text{holomorphic structures on } (E,h) \text{ whose} \\ \text{Hermitian connections are irreducible and H-E} \end{matrix} \right\} \Big/ \left\{ \begin{matrix} \text{SU(2) gauge} \\ \text{transformations} \end{matrix} \right\}, \tag{8}$$

which is isomorphic to the space of equivalent classes of irreducible ASD connections on P by Proposition 8, and the space

$$\left\{ \begin{matrix} \text{holomorphic structures on } E \text{ that} \\ \text{admit irreducible H-E connections} \end{matrix} \right\} \Big/ \left\{ \begin{matrix} \text{GL(2) gauge} \\ \text{transformations} \end{matrix} \right\}. \tag{9}$$

There is an obvious map Φ from (8) to (9) that sends $(E, h, \overline{\partial})$ to $(E, \overline{\partial})$.

Proposition 13. *The map Φ is one-to-one and onto.*

---

[2]In the literature this is also called Mumford stable or μ-stable.

Proof. We first show that Φ is one-to-one. Let $\overline{\partial}_1$ and $\overline{\partial}_2$ be two holomorphic structures having h as their H-E metrics. Assume there is a GL(2)-gauge transformation $g: E \to E$ such that $g^*(\overline{\partial}_1) = \overline{\partial}_2$. Then $g^*(h)$ is also a H-E metric of $\overline{\partial}_2$. By the uniqueness of the Hermitian-Einstein metric, there is a constant $c > 0$ such that $g^*(h) = c \cdot h$. By replacing g with $c^{-\frac{1}{2}}g$, we can assume $g^*(h) = h$. Namely, g preserves h. We claim that for some real θ the new gauge transformation $e^{i\theta}g$ preserves the volume form $\mathbf{1} \in \wedge^2 E$ as well. Indeed, because $(\wedge^2 E, \wedge^2 h)$ is a flat bundle, the (2) of Proposition 8 implies that the flat section $\mathbf{1} \in \wedge^2 E$ is holomorphic. Hence $\wedge^2 g$ must send $\mathbf{1}$ to a holomorphic section in $\wedge^2 E$, which must be $e^{-i2\theta}\mathbf{1}$ for some real θ, since $\wedge^2 h = 1$. Therefore by replacing g with $e^{i\theta}g$, g is an SU(2)-gauge transformation and $g^*(\overline{\partial}_1) = \overline{\partial}_2$. This shows that Φ is one-to-one.

Next we show that Φ is onto. Let $\overline{\partial}_0$ be any holomorphic structure on E that admits a H-E metric, say h_0. We choose a GL(2)-gauge transformation g such that $g^*(h_0) = h$. Then $\overline{\partial} = g^*(\overline{\partial}_0)$ will have $h = g^*(h_0)$ as its H-E metric. This completes the proof of the Proposition.

In algebraic geometry, the space that occurs naturally is, in our setting, the space of rank two slope-stable holomorphic vector bundles V with $c_1(V) = 0$ and $c_2(V) = d$ modulo equivalence relation. (Here by $c_1(V) = 0$ we mean $\wedge^2 V \cong \mathbf{C}_X$ is the trivial line bundle.) Here two vector bundles V_1 and V_2 are equivalent if they are isomorphic as holomorphic vector bundles. Note that as smooth vector bundles both V_1 and V_2 are isomorphic to E. Then after fixing (smooth) isomorphisms $E \cong V_1$ and $E \cong V_2$ we obtain holomorphic structures $\overline{\partial}_1$ and $\overline{\partial}_2$ on E, and the isomorphism between V_1 and V_2 is then induced by a GL(2)-gauge transformation f of E such that $\overline{\partial}_2 = f^*(\overline{\partial}_1)$. Therefore the space

$$\left\{ \begin{array}{l} \text{slope-stable holomorphic vector bundles } V \\ \text{with } \operatorname{rank} V = 2,\ c_1(V) = 0 \text{ and } c_2(V) = d \end{array} \right\} \Big/ \{\text{equivalence}\}$$

is canonically isomorphic to the space

$$\left\{ \begin{array}{c} \text{slope-stable} \\ \text{holomorphic structures on } E \end{array} \right\} \Big/ \left\{ \begin{array}{c} \text{GL(2) gauge} \\ \text{transformations} \end{array} \right\}.$$

We call the former the moduli space of rank two slope-stable vector bundles having $c_1 = 0$ and $c_2 = d$.

Combined with Theorem 12 and Proposition 13, we obtain the following theorem.

Theorem 14. *Let X be any Kähler surface. Then the moduli space of gauge equivalent classes of irreducible Anti-Self-Dual connection on an SU(2) principal bundle P with $c_2(P) = d$ is canonically isomorphic to the moduli space of rank two slope-stable vector bundles V with $c_1 = 0$ and $c_2 = d$.*

In order to attack the first part of theorem 12, we need to find an effective way to estimate the degree of sub or quotient sheaves. In our context, this can best be illustrated by using the curvature representation of Chern classes.

Let E be any holomorphic vector bundle and D be a connection with curvature $\Theta(D)$. Since $\Theta(D)$ is a section of 2-form with value in the bundle $E^\vee \otimes E$, we can take the trace of $\Theta(D)$ to get a 2-form

$$\frac{\sqrt{-1}}{2\pi}\operatorname{tr}(\Theta(D)). \tag{10}$$

According to Chern-Weil theory, this is a closed form representing the first Chern class of the bundle E. Consequently, the degree of E can be calculated by the formula

$$\deg(E) = \int_X \frac{\sqrt{-1}}{2\pi}\operatorname{tr}(\Theta(D)) \wedge \omega. \tag{11}$$

Now assume $L \subset E$ is any *subbundle*. By using the Hermitian metric h of E, we can write E as a direct sum $L \oplus L^{\perp}$, where $L^{\perp}$ is the perpendicular of $L \subset E$. ($L^{\perp}$ is a smooth subbundle of E.) Note that usually $L^{\perp} \subset E$ is not a holomorphic subbundle. We first consider the restriction of D to sections in L,

$$D_{|L} : \Omega^0(L) \longrightarrow \Omega^1(E).$$

Because $\Omega^1(E) = \Omega^1(L) \oplus \Omega^1(L^{\perp})$, we can write $D_{|L} = D_L \oplus A$, where

$$D_L : \Omega^0(L) \longrightarrow \Omega^1(L) \tag{12}$$

and

$$A : \Omega^0(L) \longrightarrow \Omega^1(L^{\perp}). \tag{13}$$

A is called the second fundamental form of $L \subset E$.

Lemma 15. *Let h_L be the restriction of h to L. Then D_L is the Hermitian connection of (L, h_L) and A is a tensor in $\Omega^{0,1}(L^{\vee} \otimes L^{\perp})$.*

Proof. We need to show that 1) D_L is a connection; 2) D_L preserves the Hermitian structure h_L and 3) $A \in \Omega^{1,0}(L^{\vee} \otimes L^{\perp})$. Here we only verify 1) and 3) and leave 2) to the readers. Indeed, for any $\xi \in \Omega^0(L)$ and smooth function f, we have

$$D(f\xi) = df \cdot \xi + fD(\xi) = df \cdot \xi + fD_L(\xi) + fA(\xi).$$

Since the left hand side is $D_L(f\xi) + A(f\xi)$, we get $D_L(f\xi) = df \cdot \xi + fD_L(\xi)$, which is the Leibnitz rule, and $A(f\xi) = fA(\xi)$, which implies A is a tensor. Further, because $L \subset E$ is a holomorphic subbundle, $D''(f\xi) \subset \Omega^{0,1}(L)$. Hence A is of type (1,0).

Similarly, by restricting D to the subbundle $L^{\perp} \subset E$, we get

$$D_{|L^{\perp}} = D_{L^{\perp}} + B. \tag{14}$$

Exercise 6. Show that $D_{L^{\perp}}$ is a Hermitian connection on $(L^{\perp}, h_{L^{\perp}})$ and B a tensor in $\Omega^{0,1}(L^{\vee} \otimes L^{\perp})$. Further, A and B are related by

$$h(A\xi, \eta) + h(\xi, B\eta) = 0, \quad \text{for } \xi \in \Omega^0(L) \text{ and } \eta \in \Omega^0(L^{\perp}). \tag{15}$$

From the above decomposition, we have

$$D = (D_L + A) + (D_{L^{\perp}} + B).$$

Written in the matrix form,

$$D = \begin{pmatrix} D_L & B \\ A & D_{L^\perp} \end{pmatrix}.$$

Hence the curvature $\Theta(D) = D \circ D$ has the form

$$\Theta(D) = \begin{pmatrix} D_L^2 + B \wedge A & D_L \circ B + B \circ D_{L^\perp} \\ A \circ D_L + D_{L^\perp} \circ A & D_{L^\perp}^2 + A \wedge B \end{pmatrix}. \tag{16}$$

Proposition 16. *Let E be a (rank 2) holomorphic bundle with H-E metric h. Then for any (rank 1) subbundle $L \subset E$, we have*

$$\deg L \leq \frac{1}{2} \deg E.$$

Furthermore, the equality holds only when both L and $L^\perp$ are holomorphic subbundles of E and $(E,h) = (L, h_{|L}) \oplus (L^\perp, h_{|L^\perp})$.

Proof. By using the curvature representation of the first Chern class, we have

$$\deg L = \int_X c_1(L) \wedge \omega = \int \frac{\sqrt{-1}}{2\pi} \operatorname{tr}(\Theta(D_L)) \wedge \omega.$$

Because $\Theta(D) \wedge \omega = c\, \omega^2 \cdot I_E$,

$$(\Theta(D_L) + B \wedge A) \wedge \omega = c\, \omega^2 \cdot I_E,$$

from (16). Hence

$$\begin{aligned} \deg L &= \int \frac{\sqrt{-1}}{2\pi} \operatorname{tr}(cI_L) \wedge \omega^2 - \int \frac{\sqrt{-1}}{2\pi} \operatorname{tr}(B \wedge A) \wedge \omega \\ &= \frac{1}{2} \int \frac{\sqrt{-1}}{2\pi} \operatorname{tr}(cI_E) \wedge \omega^2 - \int \frac{\sqrt{-1}}{2\pi} \operatorname{tr}(B \wedge A) \wedge \omega. \end{aligned} \tag{17}$$

Let A^* be the conjugate of the transpose of A with respect to the Hermitian metric h. Since A is a (1,0)-form and $A^* + B = 0$ from Exercise 6, it is straight forward to check that the form

$$-\frac{\sqrt{-1}}{2\pi} \operatorname{tr}(B \wedge A) \wedge \omega = \frac{\sqrt{-1}}{2\pi} \operatorname{tr}(A^* \wedge A)$$

is a non-positive definite (2,2)-form. Hence

$$\deg L < \frac{1}{2} \int \frac{\sqrt{-1}}{2\pi} \operatorname{tr}(cI_E) \wedge \omega^2 = \frac{1}{2} \deg E$$

unless

$$\int \frac{\sqrt{-1}}{2\pi} \operatorname{tr}(A^* \wedge A) = 0,$$

which is possible only when $A = 0$. In this case, both L and $L^\perp$ are holomorphic subbundles. Because L and $L^\perp$ are perpendicular under h, we obtain a holomorphic decomposition

$$(E,h) = (L, h_{|L}) \oplus (L^\perp, h_{|L^\perp}).$$

This completes the proof of the proposition.

This proposition says that any subbundle of a (rank 2) holomorphic vector bundle E with irreducible H-E connection has degree less than half of the degree of E. In order to show that E is slope-stable, we need to establish similar inequality for all rank 1 subsheaf $\mathcal{L} \subset \mathcal{O}(E)$. This should hold true because of the following observation: Let $\mathcal{L} \subset \mathcal{O}(E)$ be any rank 1 subsheaf. Then possibly after replacing $\mathcal{L}$ with a larger degree subsheaf, we can assume that away from a finite set $\Lambda \subset X$, $\mathcal{L} \subset \mathcal{O}(E)$ is induced by a subbundle $L_{|X-\Lambda} \subset E_{|X-\Lambda}$, where L is a line bundle on X with $\deg L = \deg \mathcal{L}$. Then over $X - \Lambda$ one can get a C^∞ splitting

$$E_{|X-\Lambda} = L_{|X-\Lambda} \oplus (L_{|X-\Lambda})^\perp,$$

using the metric h, and a decomposition of D to get

$$\begin{aligned}
&\int_{X-\Lambda} \frac{\sqrt{-1}}{2\pi} \operatorname{tr}(\Theta(D_L)) \wedge \omega \\
&\qquad = \frac{1}{2} \int_{X-\Lambda} \frac{\sqrt{-1}}{2\pi} \operatorname{tr}(cI_E) \wedge \omega^2 - \int_{X-\Lambda} \frac{\sqrt{-1}}{2\pi} \operatorname{tr}(A^* \wedge A) \wedge \omega \\
&\qquad \le \frac{1}{2} \deg E.
\end{aligned}$$

If one can show that $\int_{X-\Lambda} \operatorname{tr}(\Theta(D_L)) \wedge \omega$ is equal to the degree of L, which is not obvious because $\Theta(D_L)$ is singular at Λ, then the above inequality gives the desired inequality on degrees of $\mathcal{L}$ and E. This will prove Proposition 16 and thus the first part of Theorem 12.

Morally speaking, one expects

$$\int_{X-\Lambda} \frac{\sqrt{-1}}{2\pi} \operatorname{tr}(\Theta(D_L)) \wedge \omega = \deg L$$

since the first Chern class of a sheaf is determined away from a (complex) codimension two subset of X. However, a rigorous proof rests on a through analysis of the singularity of $\Theta(D_L)$, which can be very complicated at best. In the following we will present a rigorous proof based on Bochner technique through a series of Exercises. This proof is due to Lübke [**Lü**].

Let E be a rank two holomorphic vector bundle.

Exercise 7. Assume E has a H-E metric h, has a (holomorphic) global section $v \in H^0(E)$ and has non-positive degree, then we must have $\deg E = 0$ and (E, h) is reducible. Namely E is a direct sum of two Hermitian vector bundles. (Hint: Show that if v is a global section of E, then the (1,1)-form $\sqrt{-1}\partial\bar{\partial}h(v,v)$ is a definite form.)

Exercise 8. Assume $\mathcal{L} \subset \mathcal{O}(E)$ is a rank 1 subsheaf. Then there is a line bundle L with $\deg L = \deg \mathcal{L}$ such that $L^\vee \otimes E$ has a non-trivial global sections. (Hint: For any rank one sheaf on X, the double dual of $\mathcal{L}$ is the sheaf of holomorphic sections of a line bundle, say L. See Remark after Definition 11.)

Exercise 9. Assume E and F have H-E metrics. Then $E \otimes F$ and $E^\vee$ have H-E metrics as well.

Exercise 10. Prove part 1 of Theorem 12 by applying Exercise 7, 8 and 9.

LECTURE 3
The Existence of Hermitian-Einstein Metrics

The goal of this section is to outline the strategy of proving the existence of H-E metrics of slope-stable holomorphic bundles, following the work of Donaldson [**Do1**] and Uhlenbeck-Yau [**UY**]. We will not try to present a detailed proof, since in doing so will require many advanced topics in analysis and algebra. The interesting readers are advised to read the excellent papers of Donaldson [**Do1,2**] and Uhlenbeck-Yau [**UY**]. (An approach combining both can be found in [**Si**].)

We first set up the general framework for proving the existence theorem. For simplicity, we will work with a rank 2 slope-stable holomorphic vector bundle E with $c_1(E) = 0$. Let $\mathcal{A}$ be the space of all Hermitian metrics on E modulo the equivalence relation: *Two Hermitian metrics h_1 and h_2 are equivalent if there is a positive c such that $h_1 = c \cdot h_2$.* Then Theorem 12 says that in $\mathcal{A}$ one and only one metric is H-E. In geometric analysis there is a well-developed theory, namely variational problem, in dealing with this type of problems. We will define a functional

$$V : \mathcal{A} \longrightarrow \mathbb{R},$$

which is an integration

$$V(h) = \int_X R(h,k) \wedge \omega, \ h \in \mathcal{A}$$

of a 2-form $R(h,k)$, where $k \in \mathcal{A}$ is fixed, and show that V attains minimum at a unique point in $\mathcal{A}$. This minimum will be the H-E metric.

After Donaldson, we will define the 2-form valued functional

$$R(\cdot,\cdot) : \mathcal{A} \times \mathcal{A} \longrightarrow \Omega^{1,1}/(\operatorname{Im}\partial + \operatorname{Im}\overline{\partial})$$

to be the one satisfying the following three conditions:

1) $R(h,h) = 0$ and for any h_1, h_2 and $h_3 \in \mathcal{A}$,

$$R(h_1,h_3) = R(h_1,h_2) + R(h_2,h_3);$$

2) If h_t is a smooth one-parameter family in $\mathcal{A}$ then

$$\frac{d}{dt} R(h_t,h_0) = 2i \operatorname{tr}\big(h_t^{-1}\dot{h}_t \cdot \Theta(h_t)\big);$$

3) $\sqrt{-1}\,\overline{\partial}\partial R(h_1,h_2) = -\operatorname{tr}\big(\Theta(h_1)\big) + \operatorname{tr}\big(\Theta(h_2)\big)$.

Lemma 17. (Donaldson) *The above conditions are consistent and complete. In other words, there is a unique functional $R(\cdot,\cdot)$ satisfying these conditions.*

Proof. We will sketch the proof given in [**Do1**]. We fix a $k \in \mathcal{A}$. Since $\mathcal{A}$ is connected, we can define $R(h,k)$ to be the integration

$$\int_{t=0}^{t=1} 2i\,\mathrm{tr}\big(h_t^{-1}\dot{h}_t \cdot \Theta(h_t)\big) \tag{18}$$

along a path h_t in $\mathcal{A}$ connecting h and k. It remains to show that (18) is independent of the choice of the path. That is for any closed contour C in $\mathcal{A}$,

$$\oint_{h\in C} 2i\,\mathrm{tr}\big(h^{-1}\delta h \cdot \Theta(h)\big) \in \mathrm{Im}\,\partial + \mathrm{Im}\,\bar{\partial}. \tag{19}$$

Let θ be a differential form on $\mathcal{A}$ (with value in $\Omega^{1,1}(E^\vee \otimes E)$) defined by

$$\theta_h(\delta h) = 2i\,\mathrm{tr}\big(h^{-1}\delta h \cdot \Theta(h)\big), \quad \delta h \in T_h\mathcal{A}.$$

Then (19) is the integration of the differential form θ along C. By Stokes theorem,

$$\oint_C \theta = \int_S \mathcal{D}\theta$$

for some $S \subset \mathcal{A}$ with $\partial S = C$ (S exists since $\mathcal{A}$ is simply connected), where $\mathcal{D}$ is the exterior differential. Therefore (19) follows if

$$\mathcal{D}\theta \in \Omega^2_{\mathcal{A}}\big(\mathrm{Im}\,\partial + \mathrm{Im}\,\bar{\partial}\big). \tag{20}$$

The verification of (20) is routine and we shall omit the details here.

The next step is to investigate the critical points of the functional V. The first observation is that critical points of V are H-E metrics. Indeed, let h_0 be a critical point of V. Then for any variation h_t of h_0, we have

$$\begin{aligned} \frac{d}{dt}V(h_t)_{|t=0} &= \int_X \frac{d}{dt}R(h_t,k)_{|t=0} \wedge \omega \\ &= 2i\int_X \mathrm{tr}\big(h_t^{-1}\dot{h}_t \cdot \Theta(h_t)\big)_{|t=0} \wedge \omega \\ &= 2i\int_X \mathrm{tr}\big(h_0^{-1}\delta h_0 \cdot \Theta(h_0) \wedge \omega\big). \end{aligned} \tag{21}$$

Obviously, (21) vanishes for all possible $h_0^{-1}\delta h_0$ if and only if $\Theta(h_0)\wedge\omega = 0$, which is exactly the Hermitian-Einstein condition for h_0.

We now prove the uniqueness of the critical points of the functional V by showing that when $(E,\bar{\partial})$ is slope-stable, then the functional is strictly convex along any line $h_t = h_0\exp(ts)$, where $t \in \mathbb{R}$ and $s \in \Omega^0(E^\vee \otimes E)$ is traceless and self-adjoint with respect to h_0. Indeed, since $h_t^{-1}\dot{h}_t = s$, from the definition of $R(\cdot,\cdot)$ we have

$$\frac{d}{dt}V(h_t) = 2i\int \mathrm{tr}\big(s\Theta(h_t)\big) \wedge \omega$$

and

$$\begin{aligned}\frac{d^2}{dt^2}V(h_t) &= 2i\int \operatorname{tr}\big(s\frac{d}{dt}\Theta(h_t)\big)\wedge\omega\\ &= 2i\int \operatorname{tr}\big(s\cdot(\overline{\partial}\partial_t s)\big)\wedge\omega\\ &= 2i\int \operatorname{tr}\big(\partial_t s\wedge\overline{\partial}s\big)\wedge\omega\\ &= 2\parallel\overline{\partial}s\parallel^2_{h_t}\geq 0.\end{aligned}$$

Because $(E,\overline{\partial})$ is slope-stable, there are no traceless holomorphic sections of $E^\vee\otimes E$. Hence $\overline{\partial}s\neq 0$ and then $\|\overline{\partial}s\|^2_{h_t}>0$. This implies that V is strictly convex along the line $h_0\exp(ts)$. Finally, because any Hermitian metrics h_0 and h_1 can be connected by a line $h_0\exp(ts)$, for some s, the functional V can have at most one critical point.

The difficult part is to establish the existence of critical points. Because the functional V is convex, we will solve its gradient flow

$$\frac{d}{dt}h_t = -\operatorname{grad}V$$

and hope that it will converge to a critical point. Obviously, one necessary condition for this approach to work is that the functional V is bounded from below. This is true in our situation.

Lemma 18. (Donaldson) *If $(E,\overline{\partial})$ is a slope-stable holomorphic bundle, then the functional $V:\mathcal{A}\to\mathbb{R}$ defined above is bounded from below.*

This lemma was proved in [**Do1**] for algebraic surfaces.

The next step is to investigate the gradient flow of the functional. Let $h_t=h_0\cdot g_t$ be the solution of the gradient flow, then a direct calculation shows that

$$h_t^{-1}\frac{d}{dt}h_t = -2i\Lambda\Theta(h_t), \tag{22}$$

where $\Lambda\Theta(h_t)=(\Theta(h_t)\wedge\omega)/\omega^2$. Because the curvature of h_0 and h_t are related by

$$\Theta(h_t)=\Theta(h_0)+\overline{\partial}(g_t^{-1}\partial_0 g_t), \tag{23}$$

(22) can be rewritten as

$$\begin{cases}\frac{d}{dt}g_t = -\big(\triangle_0 g_t + i(\Lambda\Theta_0 g_t + g_t\Lambda\Theta_0) + 2i\Lambda(\overline{\partial}_0 g_t\cdot g_t^{-1}\partial_0 g_t)\big)\\ g_0=\mathrm{id}\end{cases}.$$

This is a parabolic equation. A standard technique in PDE shows that g_t exists for all t. So we are left with the task of proving that the limit

$$g_\infty = \lim_{t\to\infty} g_t$$

exists, say, in C^2.

In case the limit g_∞ does exist and satisfy some mild, say L^2, estimate, then (applying the standard technique in analysis one sees that) g_∞ is smooth and

$g_t \to g_\infty$ strongly. In such cases, $h_\infty = h_0 g_\infty$ will be a critical point of the functional. This can be seen as follows: Since h_t is a gradient flow of V,

$$\frac{d}{dt}V(h_t) = -\parallel \Lambda\Theta(h_t) \parallel^2 . \tag{24}$$

By integrating (24) over $[0,\infty)$, we get

$$\begin{aligned} -\infty &< -V(h_0) + V(h_\infty) \\ &= \int_0^\infty \frac{d}{dt}V(h_t)\,dt = -\int_0^\infty \parallel \Lambda\Theta(h_t) \parallel^2 \,dt. \end{aligned}$$

Since $\parallel \Lambda\Theta(h_t) \parallel_{L^2} \to \parallel \Lambda\Theta(h_\infty) \parallel_{L^2}$, $\Lambda\Theta(h_\infty)$ must be zero. This shows that h_∞ is H-E.

It remains to show that $\lim g_t$ exists. We will sketch the proof given in [**UY**], since their approach uses the stability condition directly. We let s_t be the self-adjoint endomorphism so that $g_t = \exp(s_t)$. Without loss of generality, we can assume h_0 satisfies the equation

$$\operatorname{tr}\bigl(\Lambda\Theta(h_0)\bigr) = 0.$$

By taking the trace of (22), we get

$$(\det h_t)^{-1}\frac{d}{dt}\det h_t = -2i\Lambda\Theta\bigl(\det h_t\bigr), \tag{25}$$

where $\det h_t$ is the induced Hermitian metric on $\wedge^2 E$. Hence by the uniqueness of solutions of heat equations, we must have

$$\det h_t = \det h_0$$

and therefore the trace of s_t vanishes identically.

Based on some delicate estimate of elliptic operator, it can be shown that if

$$\parallel s_t \parallel_{L^2} \le C$$

for some constant C independent of t, then s_t converges to a smooth s_∞, giving rise to the desired H-E metric. Now assume the otherwise. Let $\lambda_{t,i}$ be eigenvalues of s_t and let $\{e_{t,1}, e_{t,2}\}$ be orthonormal frames of E (at $x \in X$ implicitly) with respect to h_0 so that

$$s_t = \lambda_{t,1}\, e_{t,1} \otimes e_{t,1}^\vee + \lambda_{t,2}\, e_{t,2} \otimes e_{t,2}^\vee.$$

We assume $\lambda_{t,1} \ge \lambda_{t,2}$ (note $\lambda_{t,1} + \lambda_{t,2} = 0$). Since $\lim s_t$ diverges, $\parallel s_t \parallel_{L^2} \to \infty$ and then $\lambda_{t,1} \to \infty$ almost everywhere. The key observation made by Uhlenbeck and Yau is that over a dense open subset $U \subset X$,

$$e_{t,2}(x) \longrightarrow e_{\infty,2}(x)$$

and the (union of) linear subspaces $L_x \subset E_{|x}$ spanned by $e_{\infty,2}(x)$ for all $x \in U$ is a holomorphic sub-line bundle. Further this line bundle extends to a rank one subsheaf $\mathcal{L} \subset \mathcal{O}(E)$ and the degree of this subsheaf is non-negative:

$$\deg \mathcal{L} \ge 0 = \frac{1}{2}\deg E.$$

This contradicts to the assumption that E is slope-stable. Hence $\| s_t \|_{L^2}$ must be bounded uniformly and then s_t converges strongly to a smooth s_∞. This leads to the proof of the existence of the H-E metrics for slope-stable vector bundles.

In the remainder of this lecture, I shall sketch the above argument of Uhlenbeck and Yau which links the stability of E to the convergence of s_t. To make the geometry more apparent, I shall ignore all the estimate in analysis and proceed without attempting to be rigorous. For those who are interested in more details, please see [**Si, UY**].

As before, let $h_t = h_0 \cdot g_t$, $t \in [0, \infty)$, be the solution of the heat equation (22). Since V is bounded from below, there is a sequence $t_k \to \infty$ such that

$$\| \Lambda\Theta(h_{t_k}) \|^2 = -\frac{d}{dt}V(h_{t_k})$$

tend to zero. Now we estimate $\| \Lambda\Theta(h_t) \|^2$. We write

$$g_t = \sum_{i=1}^{2} e^{\lambda_{t,i}} e_{t,i} \otimes e_{t,i}^{\vee}$$

as before, which depends implicitly on $x \in X$. Let $A_{t,i}^j$ be the 1-form on X defined by

$$\partial_0 e_{t,i} = \sum_{j=1}^{2} A_{t,i}^j e_{t,j}.$$

Then

$$\partial_0 e_{t,i}^{\vee} = -\sum_{j=1}^{2} A_{t,j}^i e_{t,j}^{\vee} \quad \text{and} \quad \overline{\partial} e_{t,i} = -\sum_{j=1}^{2} \overline{A_{t,j}^i} e_{t,j},$$

since $(e_{t,1}, e_{t,2})$ are orthonormal basis of E with respect to h_0. Therefore,

$$\partial_0 g_t = \sum_{i=1}^{2} \partial_0 \lambda_{t,i}\, e^{\lambda_{t,i}} e_{t,i} \otimes e_{t,i}^{\vee} + \sum_{i,j}^{2} (e^{\lambda_{t,i}} - e^{\lambda_{t,j}}) A_{t,i}^j\, e_{t,j} \otimes e_{t,i}^{\vee}$$

and

$$\overline{\partial} g_t = \sum_{i=1}^{2} \overline{\partial} \lambda_{t,i}\, e^{\lambda_{t,i}} e_{t,i} \otimes e_{t,i}^{\vee} + \sum_{i,j}^{2} (-e^{\lambda_{t,i}} + e^{\lambda_{t,j}}) \overline{A_{t,j}^i}\, e_{t,j} \otimes e_{t,i}^{\vee}.$$

Hence,

$$\operatorname{tr}\left(\overline{\partial} g_t \wedge g_t^{-1} \partial_0 g_t\right) = \sum_{i=1}^{2} e^{\lambda_{t,i}} \overline{\partial} \lambda_{t,i} \wedge \partial_0 \lambda_{t,i} + \sum_{i,j}^{2} e^{-\lambda_{t,j}} (e^{\lambda_{t,i}} - e^{\lambda_{t,j}})^2 \overline{A_{t,i}^j} \wedge A_{t,i}^j.$$

Let $c_t = \max_X(\lambda_{t,1}, \lambda_{t,2})$. Now based on (23), we get

$$\begin{aligned}
(26) \qquad & \frac{\sqrt{-1}}{2\pi} \int_X e^{-c_t} \operatorname{tr}\left(g_t\left(\Theta(h_t) - \Theta(h_0)\right)\right) \wedge \omega \\
= & \frac{\sqrt{-1}}{2\pi} \int_X e^{-c_t} \operatorname{tr}\left(g_t\, \overline{\partial}(g_t^{-1} \partial_0 g_t)\right) \wedge \omega \\
= & \frac{\sqrt{-1}}{2\pi} \int_X e^{-c_t} \operatorname{tr}\left(g_t^{-1} \partial_0 g_t \wedge \overline{\partial} g_t\right) \wedge \omega,
\end{aligned}$$

by Stokes theorem. By the previous calculation, the above is equal to

$$-\int_X e^{-c_t}\Big(\sum_{i=1}^{2}|\partial_0\lambda_{t,i}|^2 e^{\lambda_{t,i}} + \sum_{i,j}^{2} e^{-\lambda_{t,j}}(e^{\lambda_{t,i}} - e^{\lambda_{t,j}})^2 |A^j_{t,i}|^2\Big). \tag{27}$$

Now we let $t = t_k$ with k large. Since the first line of (26) is at most

$$\| \Lambda\Theta(h_t) - \Lambda\Theta(h_0) \|^2 \cdot \| e^{-c_t} g_t \|^2,$$

which is certainly bounded, (27) will be bounded as well. Now assume $\lim h_{t_k}$ diverges, then

$$\lim_{k\to\infty} \| s_{t_k} \|_{L^2} \longrightarrow \infty.$$

If we assume $\lambda_{t_k,1}$ diverge uniformly, which is justified (to some extent) by the boundedness of

$$\int_X e^{-c_t} \sum_{i=1}^{2} e^{\lambda_{t_k,i}} |\partial_0 \lambda_{t_k,i}|^2, \tag{28}$$

then the boundedness of

$$\int_X e^{-c_t} \sum_{i,j}^{2} |A^j_{t_k,i}|^2 (e^{\lambda_{t_k,i}} - e^{\lambda_{t_k,j}})^2 e^{-\lambda_{t_k,j}}$$

implies that

$$\lim_{t\to\infty} A^2_{t_k,1} = 0,$$

since

$$e^{-c_t}\,(e^{\lambda_{t_k,1}} - e^{\lambda_{t_k,2}})^2 e^{-\lambda_{t_k,2}} \longrightarrow \infty$$

almost everywhere. Following the identity

$$\bar\partial e_{t_k,2} = \overline{A^2_{t_k,1}}\, e_{t_k,1} + \overline{A^2_{t_k,2}}\, e_{t_k,2},$$

we see that the limit $e_{\infty,2}$ of $e_{t_k,2}$ (which exists over a dense open subset of X) has the property that $\bar\partial e_{\infty,2}$ is a multiple of $e_{\infty,2}$. Namely, the line bundle spanned by $e_{\infty,2}$ will be holomorphic. What is proved in [**UY**] is that over a dense open subset $U \subset X$ with (real) codimension 4 compliment, the lines $\mathbb{C}\, e_{t_k,2} \subset E_x$ converges to $\mathbb{C}\, e_{\infty,2} \subset E_x$ and the union of the later form a line bundle on U. Further, this line bundle extends to X as a coherent subsheaf of $\mathcal{O}(E)$. We denote this subsheaf by $\mathcal{L}$.

The final step of the proof is to show that $\mathcal{L}$ is a destabilizing subsheaf of $\mathcal{O}(E)$, namely, $\deg \mathcal{L} \geq 0$. This will contradict to the assumption that E is slope-stable. For simplicity, we assume $e_{\infty,2}$ is defined everywhere on X, i.e. $\mathcal{L} \subset \mathcal{O}(E)$ is the sheaf of a subline bundle $L \subset E$. Then by a formula similar to (17),

$$\deg \mathcal{L} = \int_X \frac{\sqrt{-1}}{2\pi}\Theta(h_0)^2_2 \wedge \omega + \int_X \frac{\sqrt{-1}}{2\pi} A^*_\infty \wedge A_\infty \wedge \omega, \tag{29}$$

where A_∞ is the second fundamental form of $L \subset E$. Here for an endomorphism Φ of E, Φ_2^2 is the coefficient of $e_{\infty,2} \otimes e_{\infty,2}^\vee$. Using (23) for $t = t_k$, the above is

$$\lim_{k\to\infty} \frac{\sqrt{-1}}{2\pi} \int_X (\Theta(h_{t_k}) - \bar\partial(h_{t_k}^{-1}\partial_0 h_{t_k}))_2^2 \wedge \omega + \frac{\sqrt{-1}}{2\pi} \int_X A_\infty^* \wedge A_\infty \wedge \omega$$
$$= \lim_{k\to\infty} \frac{\sqrt{-1}}{2\pi} \int_X \left(-(\bar\partial(h_{t_k}^{-1}\partial_0 h_{t_k}))_2^2 + A_\infty^* \wedge A_\infty\right) \wedge \omega,$$

since $\| \Theta(h_{t_k}) \wedge \omega \|_{L^2} \to 0$. Now we estimate the above limit. Let $\pi_{t_k} = e_{t_k,2} \otimes e_{t_k,2}^\vee$ be the projection. Then

$$\frac{\sqrt{-1}}{2\pi} \int_X (\bar\partial(h_{t_k}^{-1}\partial_0 h_{t_k}))_2^2 \wedge \omega = \frac{\sqrt{-1}}{2\pi} \int_X \operatorname{tr}(\bar\partial(h_{t_k}^{-1}\partial_0 h_{t_k}) \circ \pi_{t_k}) \wedge \omega$$
$$= \frac{\sqrt{-1}}{2\pi} \int_X \operatorname{tr}(h_{t_k}^{-1}\partial_0 h_{t_k} \wedge \bar\partial\pi) \wedge \omega$$
$$= \frac{\sqrt{-1}}{2\pi} \int_X ((e^{\lambda_{t_k,2}-\lambda_{t_k,1}} - 1)A_{t_k,2}^1 \wedge \overline{A_{t_k,2}^1}$$
$$- (e^{\lambda_{t_k,1}-\lambda_{t_k,2}} - 1)A_{t_k,1}^2 \wedge \overline{A_{t_k,1}^2}) \wedge \omega.$$

Note that when $k \to \infty$, $A_{t_k,1}^2 \to 0$. Therefore by Fatou's lemma

$$\lim_{k\to\infty} \frac{\sqrt{-1}}{2\pi} \int_X \left(-\bar\partial(h_{t_k}^{-1}\partial_0 h_{t_k}) \circ \pi_{t_k} + A_\infty^* \wedge A_\infty\right) \wedge \omega$$
$$= \lim_{k\to\infty} \frac{\sqrt{-1}}{2\pi} \int_X \left(A_{t_k,2}^1 \wedge \overline{A_{t_k,2}^1} + e^{\lambda_{t_k,1}-\lambda_{t_k,2}} A_{t_k,1}^2 \wedge \overline{A_{t_k,1}^2} + A_\infty^* \wedge A_\infty\right) \wedge \omega$$
$$\geq \lim_{k\to\infty} \int_X \left(|A_{t_k,2}^1|^2 - |A_\infty|^2\right) \wedge \omega \geq 0,$$

since $A_{t_k,2}^1 \to A_\infty$. This shows that $\deg \mathcal{L} \geq 0$ as desired.

In closing this lecture, I will say a few words about this mysterious stability condition. The notion of slope-stable vector bundles was first introduced by Mumford for vector bundles over curves (Riemann surfaces) to ensure the existence of moduli of vector bundles. The stability can be thought of as a non-degeneracy condition on vector bundles and its exact form comes from Geometric Invariant Theory in algebra. After decades of progress, the notion of stability is much better understood in algebraic geometry. However, in light of the theorem presented in this lecture series and other similar results relating stability in algebraic geometry and existence of solutions of equations in analysis, one must ponder how an algebraic notion can possibly fit so well with that in analysis, taking into account the diverse nature of algebra and analysis. There likely is no simple answer to this question. Nevertheless, the clue may very well lies in the "principle", pointed out to the author by D. Gieseker, that algebra is a "finite dimensional approximation" of the analysis. Taking Donaldson's approach to the existence theorem of H-E connection as an example, one can see clearly how analysis is approximated by algebra and why slope-stable bundles and Hermitian-Einstein connections are two sides of the same object. The space of metrics $\mathcal{A}$ (modulo scalars) is acted on by the (complexified) gauge group $\mathcal{G}_{\mathbf{C}}$. For any $h \in \mathcal{A}$, there is an H-E metric in the orbit $\mathcal{G}_C \cdot h$ if and only if the functional V (defined in lecture 2) is proper over $\mathcal{G}_C \cdot h$. As to the Geometric Invariant Theory, the central problem is to determine all stable orbits of a (complex) reductive group G acting linearly on a linear space W. (See lecture notes of D. Gieseker in this proceeding.) If one adapts W a hermitian metric $||\cdot||^2$ preserved by a maximal compact subgroup of G, then an orbit $G \cdot w$ is stable if and only if $||\cdot||^2$ is proper over this orbit. Both cases concern orbits of a complex group acting on a space and in both cases the test condition is whether a functional is proper. This parallel suggests strongly that there should be a more direct link between Hermitian-Einstein metrics and stable vector bundles. There are other problems relating algebra and analysis that are similar to this problem. One particular interesting case is the existence of Kähler-Einstein metric on varieties of positive definite Ricci curvatures, a long standing conjecture. A recent work of G.Tian [**Ti**] points out the existence of K-E metrics is also related to the stability of varieties themselves. The exact relation in this case is still not clear and subject to future investigation.

References

[Do1] Donaldson, S.K., *Anti-self-dual yang-Mills connections over complex algebraic surfaces and stable vector bundles*, Proc. London Math. Soc. **50 No. 3** (1985), 1-26.

[Do2] Donaldson, S.K., *Infinite determinants, stable bundles and curvature*, Duke Math. J. **54 No. 1**, 231-247.

[DK] Donaldson, S.K and Kronheimer, P.B., *The geometry of four manifolds*, Oxford University Press, 1990.

[Ko] Kobayashi, S., *Differential geometry of complex vector bundles*, Iwanami Shoten, Publishers and Princeton University Press.

[Lü] Lübke, M., *Stability of Einstein-Hermitian vector bundles*, Manuscripta Math. **42**, 245-257.

[Si] Simpson, C., *Constructing variation of hodge structure using yang-Miles theory and applicaitons to uniformization*, Jour. of A. M. S. **1** (1988), 867-918.

[Ti] Tian, G., preprint.

[UY] Uhlenbeck, K. K. and Yau, S-T., *The existence of Hermitian Yang-Mills connections on stable bundles over Kähler manifolds*, Comm. Pure and Appl. Math. **39**, 257-293.

An Introduction to Gauge Theory

John W. Morgan

IAS/Park City Mathematics Series
Volume 4, 1998

An Introduction to Gauge Theory

John W. Morgan

LECTURE 1
The Context of Gauge Theory

Gauge theory is the study of principal bundles, connections on them, and the curvatures of these connections. Much of it is an analytic study of certain especially important types of connections – called (Anti)-Self-Dual connections. These objects were introduced by physicists and have proved to be a central ingredient in modern particle physics, being the mechanism that allows for unifying various of the four fundamental forces. The subject thus has many sides – the physical, differential geometric, analytic, and the topological being the most important. We are interested in the application of these techniques to the classification of four-dimensional smooth manifolds. Before turning our attention to these applications, however, we need to understand the other sides of the subject. In fact, most of this course will consist of a basic introduction to the analytic, topological, and differential geometric aspects of gauge theory. Only toward the end will we turn to its applications. Even then we will only be able to scratch the surface – a much more detailed study than we have time for is necessary before one can attempt more serious applications.

Before we begin on this long and complex study, let us set the context of the eventual applications. That is the purpose of this section.

The classical invariants of a smooth manifold are first of all its homotopy type, which is often replaced by, say, its cohomology ring. This is enough information to distinguish complex projective n space from a torus or a sphere, but is not in general sufficient information to distinguish all manifolds of a given dimension. The next level of information is the characteristic classes – the Stiefel-Whitney classes $w_i(M)$ and the Pontrjagin classes $p_i(M)$ or Chern classes $c_i(M)$ if M is a complex manifold. It turns out that one can have homotopy equivalent manifolds which are distinguished by their characteristic classes. For example, in a roundabout way these can be used to distinguish homotopy spheres, and show that

[1]Columbia University, Department of Mathematics, New York, NY.

certain topological manifolds do not admit smooth structures. It turns out that for higher dimensional manifolds (those of dimension at least 5) and, for simplicity, simply connected manifolds that these invariants are enough to determine the manifold up to finitely many diffeomorphism possibilities. The complete story is quite complicated, too complicated to formulate in complete generality, but one whose basic outlines are well-understood and where explicit complete computation in many examples is feasible. It is the story of high dimensional surgery theory, or the classification of high dimensional manifolds. As an example, there are exactly 28 smooth manifolds up to diffeomorphism which are homotopy equivalent to S^7. What is mainly relevant for us is that this beautiful and powerful theory does not hold for smooth 4-dimensional manifolds.

It is easy to understand the homotopy type of a simply connected, oriented 4-manifold B. Because of Poincaré duality there is only one interesting homology group $H_2(B) = H_2(B;\mathbf{Z})$. This group is a free abelian group of finite rank. More importantly, it has extra structure, again as a consequence of Poincaré duality. This structure is a symmetric pairing

$$H_2(B) \otimes H_2(B) \to \mathbf{Z}$$

which is unimodular (i.e., its adjoint is an isomorphism). The isomorphism type of this pairing is equivalent to the homotopy type of B. The only characteristic classes of B are $w_2(B)$ and $p_1(B)$. The first is an element in $H^2(B;\mathbf{Z}/2\mathbf{Z})$ and the second is an element in $H^4(B;\mathbf{Z})$ and hence can be thought of as an integer (since B is connected and oriented). Each of these classes is determined by the pairing on $H_2(B)$. The second Stiefel-Whitney class is simply the class that measures $x \cdot x$ (mod 2) whereas by the Hirzebruch Index Theorem $p_1(B)$ is identified with three times the signature of the symmetric pairing on $H_2(B)$ (see Problem 8). This means that all the classical invariants of a four-manifold B are determined by the intersection pairing on $H_2(B)$.

The differential topological constructions which work so well in higher dimensions can be used to show that if M and N are simply connected four-manifolds with isomorphic intersection pairings then $M \times S^1$ and $N \times S^1$ are diffeomorphic by a diffeomorphism homotopic to $f \times \mathrm{Id}_{S^1}$ for any given homotopy equivalence f between M and N (or what amounts to the same thing, any isomorphism between $H_2(M)$ and $H_2(N)$ which is an isomorphism of spaces with pairings). Unfortunately, (or rather fortunately) these techniques do not apply directly to N and M, and before the advent of gauge theory it was a complete mystery as to whether or when M and N were diffeomorphic.

Gauge theory can be used to produce new algebraic invariants of smooth four-manifolds. These invariants can be considered in some sense to be non-linear versions of the Pontrjagin classes, since they are derived from non-linear constructions involving bundles over the four-manifold. These invariants have no analogue in the higher dimensions (or in dimension three for that matter). They are peculiarly four-dimensional invariants. Using these invariants we now have a much clearer picture of when N and M are diffeomorphic, though by no means do we have a complete solution to the classification problem. In favorable circumstances we can compute or partially compute these invariants and then show that certain examples of homotopy equivalent four-manifolds are not diffeomorphic since these invariants differ. To complete the picture we would need geometric constructions to show that when these invariants agree the manifolds are diffeomorphic. In many examples this is indeed the case but there is no conjecture as to what type of geometric constructions might yield such a result, or even if it is reasonable to conjecture that such a result is true.

Some of the typical questions that were asked but unanswered before 1980 include:

Which symmetric bilinear pairings are realized as the intersection pairings of smooth four-manifolds?

Does the $K3$ surface, whose intersection pairing is the direct sum of three hyperbolic pairs and a negative definite form, split off one, two, or three copies of $S^2 \times S^2$ reflecting the algebraic decomposition of its form?

There are algebraic surfaces (certain elliptic surfaces) whose intersection forms agree with that of the $K3$-surface, but whose complex structure is very different from that of the $K3$ (other elliptic surfaces). Are these surfaces diffeomorphic to the $K3$ surface?

Is every simply connected four-manifold a connected sum of algebraic surfaces?

Thanks to gauge theory we have a much better understanding of the topology of smooth four-manifolds. For example, we now have complete answers to all these questions except the first one, where we have a partial answer as well as a conjectured complete answer.

1.1. Problems

1. State and prove Poincaré duality for closed smooth, oriented n-dimensional manifolds. Show that H_2 of a simply connected 4-manifold is a free abelian group.
2. Let M be a closed oriented n-manifold. Define the intersection form
$$q_M\colon H_i(M) \otimes H_{n-i}(M) \to \mathbf{Z}$$
by
$$q_M(x,y) = \langle PD^{-1}(x), y\rangle,$$
where PD is the Poincaré duality isomorphism.
3. Show that the adjoint of q_M is an isomorphism from
$$H_i(M)/\text{Torsion} \to \text{Hom}(H_{n-i}(M), \mathbf{Z}).$$
4. Show that
$$q_M(x,y) = \langle PD^{-1}(x) \cup PD^{-1}(y), [M]\rangle.$$
5. Let M be a closed smooth oriented manifold of dimension $2n$. Consider the restriction of the intersection pairing q_M to $H_n(M)/\text{Torsion}$. If we choose a basis for $H_n(M)/\text{Torsion}$ then the intersection pairing is represented by a $(-1)^n$ symmetric matrix whose determinant is ± 1 (a so-called unimodular matrix). (Such pairings are called *unimodular*, or *perfect*.) Show that changing the basis for this group replaces a matrix Q by a matrix of the form $A^{tr}QA$ for A a unimodular matrix.
6. Show that for any symmetric, unimodular bilinear form on a lattice Λ, there are vectors $c \in \Lambda$ with the property that for any vector v,
$$< v,c > = < v,v > \pmod 2.$$

(A lattice is a finitely generated free abelian group.) Such a vector is called a *characteristic* vector. Show that a characteristic vector is unique (mod 2).

7. Let $\langle , \cdot , \rangle$ be a symmetric, unimodular pairing on a lattice Λ. Show that all characteristic vectors in Λ have the same square modulo 8 and that they are all equivalent modulo two. Show that, conversely, if an equivalence class modulo 2 in Λ has the property that any vector in the equivalence class has some given square modulo 8, then that equivalence class is the equivalence class of characteristic vectors.
8. Show that any symmetric bilinear pairing on a finite dimensional real vector space V. can be diagonalized with all diagonal entries ± 1. Show that the number of $+1$'s (hence -1's) on the diagonal is an invariant of the pairing. The difference between the number of $+1$'s and the number of -1's is called the *signature* of the bilinear form. A form is said to be *positive definite*. resp., *negative definite* if $x \cdot x > 0$, resp., $x \cdot x < 0$, for all $x \neq 0$ in V. Show that for a positive definite pairing the signature is equal to the rank and for a negative definite pairing the signature is equal to minus the rank.

9*. Show that for a unimodular form on a lattice the signature reduces modulo 8 to the square of a characteristic element.

10. Show that for a four-manifold M with no 2-torsion, the intersection form determines $w_2(TM)$ in the sense that for an $v \in H^2(M)$, $< v, v >= < w_2(TM), v >$; i.e. $w_2(TM)$ is characteristic.
11. Show that there is a well-defined homomorphism

$$H_n(M) \to \mathbf{Z}/2\mathbf{Z}$$

given by sending $x \in H_n(M)$ to $x \cdot x$ (mod 2). Dual to this pairing is an element $v_n(M) \in H^n(M; \mathbf{Z}/2\mathbf{Z})$ called the nth Wu class.

12. Assuming that $H_{n-1}(M)$ has no 2-torsion, show that $v_n(M)$ vanishes if and only if there is a matrix representative for the intersection pairing so that all the diagonal entries are even if and only if every matrix representative for the intersection pairing has even diagonal entries.
13. Define the symmetric unimodular pairing associated with the Dynkin diagram of the Lie algebra E_8. This is an even, unimodular, negative definite pairing on a lattice of rank eight.
14. Let (L, q) be a positive definite, symmetric, unimodular pairing with L a lattice. Show that the number of vectors $\ell \in L$ with $q(\ell, \ell) = 1$ is finite and is at most twice the rank of L. Show that if the number of such vectors is exactly equal to twice the rank of L, then L is diagonalizable over the integers.
15. Show that $E_8 \oplus < -1 >$ is not diagonalizable over the integers.
16. Suppose that $L \subset \mathbf{R}^n$ is a lattice (i.e., L is a free abelian group of rank n which spans $\mathbf{R}^n$). Show that the volume of the quotient torus $\mathbf{R}^n/L$ with the induced flat metric from $\mathbf{R}^n$ is equal to $\sqrt{det(\langle e_i, e_j \rangle)}$, where $\{e_i\}_{i=1}^n$ is any basis for L. Show that if $r > 0$ has the property that the ball of radius r in $\mathbf{R}^n$ has volume larger than that of the quotient torus, then there is a point of $L - \{0\}$ within distance $2r$ of 0.
17. Show any integral lattice of determinant ± 1 in $\mathbf{R}^n$ with $n \leq 4$ has a vector of length 1.

18. Show that any unimodular positive definite symmetric pairing of rank at most 4 is diagonalizable (i.e., represented *over the integers*) by a diagonal matrix with ones down the diagonal).
19. Show that any unimodular indefinite form of rank 2 has a non-zero vector with self-intersection equal to zero.
20. Classify indefinite unimodular forms of rank 2.
21*. Using 18 classify unimodular integral forms of rank ≤ 4.
22. It is a theorem that any symmetric bilinear integral form q of rank at least 5 which is indefinite has a non-zero vector x with $q(x, x) = 0$. Using this and Problem 17, establish the classification of odd indefinite forms.
23. What are the intersection forms of $\mathbf{C}P^2$, $S^2 \times S^2$, and S^4?
24. Take a complex line bundle $\mathcal{L} \to S^2$ over the 2-sphere with c_1 equal to n. Form the projective bundle of $\mathcal{L} \oplus \mathbf{C} \to S^2$. Show that for n even the intersection form for this manifold is isomorphic to the form for $S^2 \times S^2$. Compute the intersection form when n is odd. Which of these manifolds is diffeomorphic to $S^2 \times S^2$?
25. Show that if M is diffeomorphic to a connected sum $A \# B$ then the intersection form of M is isomorphic to the direct sum of those of A and B.

LECTURE 2
Principal Bundles and Connections

Principal bundles and connections – the basics

2.1. Principal bundles

Among other things, a principal bundle has a structure group G, which is a Lie group, and a base B which is a topological space. Almost always we work in one of the following two situations. One possibility is that we assume that the base is a simplicial complex (or a CW complex). This is convenient for doing homotopy theory. The other possibility is that we assume that the base is a smooth manifold, in which case we assume that the principal bundle is a smooth bundle so that the total space is also a smooth manifold.

Notice that on the product $B \times G$ there is a natural right action of G by right multiplication on the second factor. This is a free action, and the quotient is identified with B. In fact the quotient map is naturally identified with the projection onto the first factor. This is an example of a principal bundle, called the *trivial* or *product* bundle with base B. In general, a principal bundle is a twisted version of this situation.

Definition 2.1.1. A *(right) principal G-bundle* consists of a triple (P, B, π) where $\pi\colon P \to B$ is a map and a continuous, free right action $P \times G \to P$ with respect to which π is invariant and so that π induces a homeomorphism between the quotient space of this action and B. Furthermore, there is an open covering $\{U_\alpha\}$ of B over which all the above data are isomorphic to the product data. That is to say for each α there exists a commutative diagram

$$\begin{array}{ccc} \pi^{-1}(U_\alpha) & \xrightarrow{\varphi_\alpha} & U_\alpha \times G \\ {\scriptstyle\pi}\downarrow & & \downarrow{\scriptstyle p_1} \\ U_\alpha & \xrightarrow{=} & U_\alpha \end{array}$$

where φ_α is a homeomorphism which is equivariant with respect to the right G-actions and p_1 is the projection onto the first factor.

The space P is called the *total space* of the principal bundle, π is called *the projection* and B is called the *base*. The maps φ_α are *called local trivializations.* Lastly, G is called the *structure group of the bundle*. If P and B are smooth manifolds, if the action of G on P is a smooth action and if π is a smooth submersion, then the principal bundle is said to be a *smooth principal bundle*. In this case it follows automatically, that one can choose the local trivializations so that the φ_α are diffeomorphisms (See Problem 5).

An *isomorphism* of G-bundles with the same base is a homeomorphism between their total spaces which is G-equivariant and which commutes with the projections to the base. A map between G-bundles over possibly different bases is a G-equivariant map between the total spaces. Such a map must be an isomorphism on each fiber and it induces a map between the base spaces.

2.2. Examples of principal bundles

Let M be a smooth manifold. Let E be the frame space for the tangent bundle of M. A point of E consists of a point $p \in M$ and a basis $\{v_1, \dots, v_n\}$ for the tangent space TM_p to M at p. The topology, and indeed the smooth structure, of E is induced in the obvious way from that of the tangent bundle TM. There is the obvious projection of E to M and an obvious action of $GL(n, \mathbf{R})$ on E. The action of $A = (a_{ij}) \in GL(n, \mathbf{R})$ on the point $(x, \{v_1, \dots, v_n\})$ gives the point $(x, \{w_1, \dots, w_n\})$ where

$$w_j = \sum_{i=1}^{n} a_{ij} v_i.$$

That is to say, the matrix A acts on the basis to produce a new basis for the same space; the expression for the new basis in terms of the old basis is given by the columns of the matrix A. It is easy to see that this defines a right action of $GL(n, \mathbf{R})$ on E.

A local coordinate chart for $U \subset M$ determines a local trivialization of this bundle in the following way. Let $U \subset M$ be an open subset with coordinates $(x_1, \dots, x_n)$. Then we define an isomorphism

$$U \times GL(n, \mathbf{R}) \to E|_U$$

$$(u, A) \mapsto \left(u, \{\frac{\partial}{\partial x_1}, \dots, \frac{\partial}{\partial x_n}\} \cdot A\right).$$

There are other related examples. Suppose that M is a riemannian n-dimensional manifold. Then we can take the space of orthonormal bases for the tangent spaces of M. This forms the total space of a principal $O(n)$-bundle. If in addition M is oriented then we can restrict to special orthogonal frames and get a principal $SO(n)$-bundle related to the tangent bundle.

All of these examples are derived in one way or another from the tangent bundle of a manifold. But not all principal bundles arise in this way. As an example, we have the tautological principal S^1-bundle over complex projective n-space. Let $\mathbf{C}P^n$ be the manifold of complex lines in $\mathbf{C}^{n+1}$. Consider S^{2n+1} as the unit sphere in $\mathbf{C}^{n+1}$. There is the obvious smooth map $S^{2n+1} \to \mathbf{C}P^n$. Also there is the natural action of the complex numbers of unit length by scalar multiplication on S^{2n+1}. These data determine a principal S^1-bundle, the tautological bundle over $\mathbf{C}P^n$.

If $\mathcal{L} \to B$ is a complex line bundle, then by choosing a hermitian inner product on $\mathcal{L}$ we can obtain the associated principal S^1-bundle consisting of the subspace of all vectors in $\mathcal{L}$ of unit length.

2.3. The transition functions

Let $\pi\colon P \to B$ be a principal G-bundle and suppose that we have an open covering $\{U_\alpha\}$ of B and local trivializations

$$\varphi_\alpha\colon \pi^{-1}U_\alpha \xrightarrow{\cong} U_\alpha \times G$$

given by, say,

$$\varphi_\alpha(e) = (\pi(e), g_\alpha(e)).$$

Associated to these data are *transition functions*

$$g_{\alpha,\beta}\colon U_\alpha \cap U_\beta \to G$$

defined by the equation

$$g_\alpha(e) = g_{\alpha,\beta}(\pi(e)) \cdot g_\beta(e)$$

for all $e \in \pi^{-1}(U_\alpha \cap U_\beta)$. Clearly, these transition functions satisfy the *cocycle condition*:

$$g_{\alpha,\beta}(x) \cdot g_{\beta,\gamma}(x) = g_{\alpha,\gamma}(x)$$

for all $x \in U_\alpha \cap U_\beta \cap U_\gamma$. It follows that $g_{\alpha,\alpha}(x) = 1$ for all $x \in U_\alpha$ and $g_{\alpha,\beta}(x) = g_{\beta,\alpha}(x)^{-1}$ for all $x \in U_\alpha \cap U_\beta$.

If φ'_α is another set of trivializations over the U_α resulting in functions $g'_\alpha\colon U_\alpha \to G$ then there are a maps $h_\alpha\colon U_\alpha \to G$ such that

$$g'_\alpha(e) = h_\alpha(\pi(e)) \cdot g_\alpha(e).$$

In this case the new transition functions are related to the old ones by

$$g'_{\alpha,\beta}(x) = h_\alpha(x) \cdot g_{\alpha,\beta}(x) \cdot h_\beta(x)^{-1}$$

for all $x \in U_\alpha \cap U_\beta$. In this case we say that the two transition cocycles differ by a *coboundary*.

The transition function language gives us another way to describe principal G-bundles over a base B. Suppose that we have an open covering $\{U_\alpha\}$ of B and functions

$$g_{\alpha,\beta}\colon U_\alpha \cap U_\beta \to G$$

which satisfy the cocycle conditions. We form a space P by taking the quotient space of

$$\coprod_\alpha U_\alpha \times G$$

under the equivalence relation $(u, g) \in U_\alpha \times G$ is identified with $(u, g_{\beta,\alpha}(u) \cdot g)$ in $U_\beta \times G$ for any $u \in U_\alpha \cap U_\beta$. (This corresponds to the equation $g_\alpha(u) = g_{\alpha,\beta}(u) \cdot g_\beta(u)$.) The cocycle condition implies that this is an equivalence relation. There is an obvious map $P \to B$ induced by the projections to the first factors on the $U_\alpha \times G$. There is also an obvious right action of G on P induced from the right multiplications of G on the $U_\alpha \times G$. It is easy to establish that these data form a principal G-bundle. If B is a smooth manifold and if the $g_{\alpha,\beta}$ are smooth maps then this construction determines a smooth principal G-bundle.

In fact, for locally contractable paracompact spaces one can identify the isomorphism classes of principal G-bundles with the space of cocycles modulo coboundaries for a fixed covering of the base space by contractable, open sets. This quotient is denoted by $H^1(B;G)$, and is called the *first Cech cohomology group* with coefficients in G.

2.4. Pullback bundles

Suppose that $\pi\colon P \to B$ is a principal G-bundle and that $f\colon A \to B$ is a continuous map. Then we can form the *pullback* $f^*P \to A$ which is a principal G-bundle. The total space of f^*P is the fibered product of the following diagram:

$$\begin{array}{ccc} & & P \\ & & \big\downarrow{\scriptstyle \pi} \\ A & \xrightarrow{\ f\ } & B \end{array}$$

This means that $f^*P \subset A \times P$ is the set of pairs $\{(a,p)|f(a) = \pi(p)\}$. The projection mapping from f^*P to A is the obvious mapping from the fibered product to A. The action of G on the total space is induced from the action of G on P. If $P \to B$ is a smooth and if f is a smooth map then the pullback is also a smooth principal bundle.

Suppose that $Q \to A$ and $P \to B$ are principal G-bundles. A principal bundle map $F\colon Q \to P$, i.e., a G-equivariant map from Q to P, induces a map $f\colon A \to B$ and an isomorphism of principal bundles $Q \to f^*P$.

2.5. Associated bundles

Suppose that $P \to B$ is a principal G-bundle and that we have a left action of G on a space F. Then we can define the associated locally trivial fiber bundle over B with fiber F by forming $P \times_G F \to B$. As an example, if G is a linear Lie group, say, G is embedded as a subgroup of $GL(n, \mathbf{R})$ for some n, then we have a natural left action of G on $\mathbf{R}^n$. Thus, associated to any principal G-bundle there is an n-dimensional real vector bundle. There are similar statements for subgroups of $GL(n, \mathbf{C})$. Of course, constructing the frame bundle of a vector bundle reverse this process, it constructs a principal bundle with structure group $GL(n, \mathbf{R})$ whose associated vector bundle is the given bundle.

Along the same lines, suppose that we have a linear representation $\rho\colon G \to Gl(V)$. Then associated to a principal G-bundle $P \to B$ and ρ there is a vector bundle

$$P \times_G V \to B.$$

As an example of this, consider the adjoint representation of G on its Lie algebra $\mathcal{G}$

$$G \times \mathcal{G} \to \mathcal{G}$$

given by

$$(g, X) \mapsto g \cdot X \cdot g^{-1}.$$

The vector bundle associated to $P \to B$ and this representation is denoted ad$P$ and is called the *adjoint bundle* of P.

More generally, suppose that $p\colon P \to B$ is a locally trivial fiber bundle with fiber F. In a generalized sense this bundle is associated to a principal bundle with structure group the group Homeo(F) of all homeomorphisms of F. (Of course, this group is not a Lie group since it is not finite dimensional.) Nevertheless, if we can reduce the structure group from the group of all homeomorphisms of F to a Lie group $G \subset \text{Homeo}(F)$, then we can find a principal G-bundle with the given fiber bundle as associated bundle. To reduce the structure group in this case means simply that we can find an open cover $\{U_\alpha\}$ of B and local trivializations $\varphi_\alpha \colon p^{-1}(U_\alpha) \to U_\alpha \times F$ so that the transition functions $g_{\alpha,\beta} \colon U_\alpha \cap U_\beta \to \text{Homeo}(F)$ have image contained in G.

2.6. Universal bundles

Consider the grassmannian $Gr(n,k)$ or n-planes in $\mathbf{R}^{n+k}$. There is a natural n-dimensional vector bundle $\xi(n,k) \to Gr(n,k)$ called the tautological bundle. Its fiber over a plane $Q \in Gr(n,k)$ is the n-plane Q itself (see Problem 8). We have the obvious inclusion $Gr(n,k) \subset Gr(n,k+1)$. We let $Gr(n,\infty)$ be the union of these spaces over all $k \geq 0$. This space can be viewed as the grassmannian of n planes in $\mathbf{R}^\infty$. There is a tautological n-plane bundle $\xi(n) \to Gr(n,\infty)$ which restricts to each $Gr(n,k)$ to give the tautological bundle $\xi(n,k)$. We let $P(\xi_n) \to Gr(n,\infty)$ be the associated principal $Gl(n,\mathbf{R})$ bundle.

Suppose that $V \to B$ is an n-dimensional vector bundle over a paracompact base B. Let $\{U_\alpha\}$ be an open covering (supposed countable for simplicity) of B such that for each α the restriction $V|_{U_\alpha}$ is isomorphic to the trivial bundle. Let $\varphi_\alpha \colon V|_{U_\alpha} \to U_\alpha \times \mathbf{R}^n$ be a trivialization. Let λ_α be a partition of unity subordinate to this covering. We form $\mu_\alpha = \lambda_\alpha \cdot (p_2 \circ \varphi_\alpha) \colon V \to \mathbf{R}^n_\alpha$. The sum

$$\mu = \sum_\alpha \mu_\alpha$$

determines a well-defined map from V to $\oplus_\alpha \mathbf{R}^n_\alpha$ which embeds each fiber of V linearly into the linear subspace with only finitely many non-zero coordinates, $\mathbf{R}^\infty$. Thus, μ defines a map $f \colon B \to Gr(n,\infty)$ and an isomorphism between V and the pullback $f^*\xi(n)$. Of course, if $P(V)$ is the principal $GL(n,\mathbf{R})$-bundle which is the frame bundle of V, then $f^*P(\xi(n))$ is identified with $P(V)$.

There is an important uniqueness statement along these lines as well. It is based on the following lemma.

Lemma 2.6.1. *Let B be a paracompact base. If $P \to B \times I$ is a principal G-bundle, then there is a G-bundle isomorphism from*

$$\left(P|_{B\times\{1\}} \times I\right) \to P.$$

Proof. First, we construct an open covering of $\{U_\alpha\}$ of B such that $P|_{U_\alpha \times I}$ is trivial for all α. This is easily done using the compactness of I and the local triviality of P. We fix such a covering and fix functions $\lambda_\alpha \colon B \to [0,1]$ with the property that the support of λ_α is contained in U_α, the support of the λ_α form a locally finite collection of closed subsets, and $\max_\alpha \lambda_\alpha(x) = 1$ for all $x \in B$. We define maps

$$r_\alpha \colon B \times I \to B \times I$$

by $r_\alpha(x,t) = (x, \max(t, \lambda_\alpha(x)))$. Notice that outside of the support of λ_α times I, r_α is equal to the identity.

We shall construct maps $R_\alpha \colon P \to P$ covering the r_α and commuting with the G-actions, and equal to the identity outside the preimage under π of the product of the support of λ_α with I. To do this we need only work in $p^{-1}(U_\alpha \times I)$, and then extend by the identity outside. The construction over $U_\alpha \times I$ is immediate from the trivialization of P over this subset.

Now we take a total ordering of the index set α and compose the R_α in the order indicated by the ordering on the index set. This produces an infinite composition in which locally all but finitely many of the maps are the identity. The resulting map covers the composition of the r_α which is the projection of $B \times I$ to $B \times \{1\}$. Such a map induces an isomorphism between the given principal G-bundles. □

Corollary 2.6.2. *If $P \to B \times I$ is a principal bundle then there is an isomorphism of principal bundles*

$$P|_{B\times\{0\}} \to P_{B\times\{1\}}.$$

It follows that if $f_0, f_1 \colon B \to Gr(n,\infty)$ are homotopic mappings then the pullbacks $f_0^* P(\xi_n)$ and $f_1^* P(\xi_n)$ are isomorphic. That is to say there is a well-defined function from the homotopy classes of maps from B to $Gr(n,\infty)$, denoted $[B, Gr(n,\infty)]$, to the set of isomorphisms classes of principal $Gl(n,\mathbf{R})$ bundles over B. We have seen that this function is onto. It remains to show that it is one-to-one. That is to say, we need to see that if $f_0, f_1 \colon B \to Gr(n,\infty)$ are maps and if the pullbacks $f_0^* P(\xi_n)$ and $f_1^* P(\xi_n)$ are isomorphic then f_0 and f_1 are homotopic.

We view the problem this way – we have two maps of $F_0, F_1 \colon P \to P(\xi_n)$ which commute with the $Gl(n,\mathbf{R})$-actions, covering the maps $f_0, f_1 \colon B \to Gr(n,\infty)$. We wish to construct a homotopy between f_0 and f_1. Let us first do a special case.

Suppose that for each $b \in B$ the planes $f_0(b)$ and $f_1(b)$ meet only in $\{0\}$. Then we form the maps $F_t \colon P \to P(\xi_n)$. Recall that a point of $P(\xi_n)$ is an n-frame in $\mathbf{R}^\infty$. If $F_0(p) = \{a_1, \ldots, a_n\}$ and $F_1(p) = \{b_1, \ldots, b_n\}$, then we define

$$F_t(p) = \{(1-t)a_1 + tb_1, \ldots, (1-t)a_n + b_n\}.$$

Clearly, under our assumption on $f_0(b)$ and $f_1(b)$, each $F_t(p)$ is indeed an n-frame in $\mathbf{R}^\infty$. This defines the required homotopy.

To complete the proof of injectivity we need to see that any pair f_0, f_1 can be deformed by homotopy until the above independence condition is satisfied. Let $L_t \colon \mathbf{R}^\infty \to \mathbf{R}^\infty$ be the homotopy defined by

$$L_t(x_1, x_2, \ldots) = (1-t)(x_1, x_2, \ldots) + t(0, x_1, 0, x_2, 0, \ldots).$$

Clearly $L_0 = \mathrm{Id}$, and L_t is a linear embedding of $\mathbf{R}^\infty$ into itself for all $t \le 1$. We can apply this homotopy to produce a homotopy

$$g_t \colon Gr(n,\infty) \to Gr(n,\infty)$$

from the identity to a map g_1 whose image is contained in the set of points which represent planes contained in the subspace with only the even coordinates non-zero.

There is an analogous homotopy h_t from the identity to a map whose image is contains in the set of points which represent planes contained in the subspace with only odd coordinates non-zero. Now given maps $f_0, f_1 \colon B \to Gr(n,\infty)$, we have that $k_0 = g_1 \circ f_0$ is homotopic to f_0 and $k_1 = h_1 \circ f_1$ is homotopic to f_1. Clearly, for every $b \in B$ the planes $k_0(b)$ and $k_1(b)$ are linearly independent, since one is contained in the "even coordinate" subspace and the other is contained in the "odd coordinate" subspace. Applying the previous corollary we see that if $f_0^* P(\xi_n)$ is isomorphic to $f_1^* P(\xi_n)$ then k_0 and k_1 are homotopic, and hence f_0 and f_1 are homotopic.

Corollary 2.6.3. *Let B be a paracompact space with a countable base for its topology. Pulling back the universal bundle $P(\xi_n)$ induces a bijective function from the set of homotopy classes of maps $[B, Gr(n,\infty)]$ to the set of isomorphism classes of principal $Gl(n,\mathbf{R})$ bundles over B, or equivalently the set of isomorphism classes of n-dimensional real vector bundles (see Problem 7).*

2.7. Connections on smooth principal bundles

Let $\pi\colon P \to B$ be a smooth principal G-bundle over an n-dimensional manifold. A connection for this bundle is an infinitesimal version of an equivariant family of cross sections. It is an n-dimensional distribution $\mathcal{H}$ (i.e., smooth family of n-dimensional linear subspaces of the tangent bundle TP of P) which is *horizontal* in the sense that the restriction of $D\pi$ to each plane in the distribution is an isomorphism onto the corresponding tangent plane to B and which is invariant under the G-action. Such a distribution is a family of complementary subspaces to the subbundle of tangents along the fibers TP^v (also called *vertical tangents*) and hence induces an isomorphism $TP_p \equiv TP_p^v \oplus TB_{\pi(p)}$.

Lemma 2.7.1. *Suppose that Γ is a connection for $P \to B$. Let $\gamma\colon [0,1] \to B$ be a smooth path and $e \in \pi^{-1}(\gamma(0))$. Then there is a unique path $\tilde{\gamma}\colon [0,1] \to P$ such that $\tilde{\gamma}(0) = e$, $\pi \circ \tilde{\gamma} = \gamma$ and $\tilde{\gamma}'(t)$ is contained in the horizontal space $\mathcal{H}_{\tilde{\gamma}(t)}$.*

Proof. To prove this one simply pulls the bundle and connection back via γ. This allows us to treat the case where B is the interval and γ is the identity map. The connection determines a vector field on the total space of the principal bundle which projects down to the vector field d/dt on the interval. The horizontal lift $\tilde{\gamma}$ that we need to construct is simply the integral path for this vector field with the given initial condition. The existence and uniqueness of the horizontal lift follows immediately from the corresponding statements for integral curves for vector fields. □

Notice that the horizontal lift $\tilde{\gamma}$ depends in an equivariant fashion on the initial point e in the sense that the horizontal lift beginning at $e \cdot g$ is simply $\tilde{\gamma} \cdot g$. Thus, we have the following corollary.

Corollary 2.7.2. *Given a smooth curve in the base $\gamma\colon [0,1] \to B$ from b_0 to b_1 a connection determines an isomorphism between the fibers $\pi^{-1}(b_0) \to \pi^{-1}(b_1)$ which is equivariant with respect to the G-actions on these fibers.*

This is the reason for the name connection – a connection gives a manner to connect distinct fibers, albeit one needs a path in the base between the image points in the base.

2.8. The differential form description of a connection

Let $\mathcal{G}$ be the Lie algebra for G. Then there is a unique one-form $\omega_{MC} \in \Omega^1(G;\mathcal{G})$ which is invariant under left multiplication by G and whose value at the identity element of G is the identity linear map from $TG_e \to \mathcal{G}$. This form is called the *Maurer-Cartan form*. It is often denoted $g^{-1}dg$. Its value on a tangent vector $\tau \in TG_g$ is equal to $g^{-1} \cdot \tau \in TG_e = \mathcal{G}$.

Lemma 2.8.1. *A connection on a smooth principal bundle $\pi\colon P \to B$ is equivalent to a differential one form $\omega \in \Omega^1(P;\mathcal{G})$ with the following properties:*

- *Under right multiplication by G the form ω transforms via the adjoint representation of G on $\mathcal{G}$; i.e.,*

$$\omega_{pg}(\tau \cdot g) = g^{-1}\omega_p(\tau) \cdot g$$

for any $p \in P$, any $\tau \in TP_p$ and any $g \in G$.
- *For any $p \in P$ consider the embedding $R_p\colon G \to P$ given by $R_p(g) = p \cdot g$. Then the pullback $R_p^*(\omega) = \omega_{MC}$.*

Proof. Suppose that we are given a form ω with the two properties. Let $\mathcal{H}_e$ be the kernel of the linear map $\omega_p\colon TP_p \to \mathcal{G}$. According to the second property that ω is required to satisfy, its restriction to the vertical tangent vectors induces an isomorphism $TP_p^v \to \mathcal{G}$. Thus, the $\mathcal{H}_p$ form a smooth distribution which projections isomorphically onto the tangent spaces to B. The only other thing to check about this distribution in order to show that it is a connection is G-invariance. But by the first property of ω it is clear that the kernels of ω are G-invariant.

Conversely, given a connection we define $\omega_p\colon TP_p \to \mathcal{G}$ to be the following composition:

$$TP_p \to TP_p^v \xrightarrow{(R_p^*)^{-1}} \mathcal{G}$$

where the first map is the linear projection with kernel $\mathcal{H}_p$. One checks easily that the resulting form satisfies the two conditions (see Problem 27). It is apparent that these constructions are inverses of each other. □

2.9. Existence of connections

We have not yet shown that principal bundles have connections. That is the subject of this paragraph.

Lemma 2.9.1. *The trivial bundle has a connection.*

Proof. The product structure $B \times G$ gives us a natural horizontal distribution, namely the tangent spaces to the first factors. □

Lemma 2.9.2. *If P has a connection, then the space of all connections on P is an affine space; the underlying vector space is $\Omega^1(B; \mathrm{ad}P)$, the space of one-forms on B with values in the vector bundle* ad$P$.

Proof. We take the point of view that a connection is a one-form on P with certain properties. It is clear that the subspace of forms satisfying these properties is an affine subspace of $\Omega^1(P;\mathcal{G})$. The difference of two connection one-forms is a one form $\delta \in \Omega^1(P;\mathcal{G})$ which transforms under the adjoint representation when transported by $g \in G$ and which vanishes on all vertical tangent vectors. It is an easy exercise (see Problem 19) to identify this space of forms on P with the pullback via π of one-forms in $\Omega^1(B; \mathrm{ad}P)$. □

Proposition 2.9.3. *Any smooth principal bundle $P \to B$ has a connection A. The space of all connections is an affine space whose underlying vector space is identified with $\Omega^1(B; \mathrm{ad}P)$.*

Proof. We cover the base B by open sets $\{U_\alpha\}$ over which P is trivial. For each α we have a connection A_α for $P|_{U_\alpha}$. Let $\{\lambda_\alpha\}$ be a partition of unity subordinate to the covering $\{U_\alpha\}$. We form

$$A = \sum_\alpha \lambda_\alpha A_\alpha.$$

This is a one-form on P with values in $\mathcal{G}$. Near the preimage of any point $b \in B$, this one-form is an affine combination of connection one-forms and hence is a connection one-form. But if that is true in the neighborhood of the preimage of each point of B then it means that A is a connection one-form on all of P.

The second statement follows from the previous lemma. □

2.10. Covariant differentiation

Suppose that A is a connection on a principal bundle $\pi\colon P \to B$, and suppose that $W \to B$ is a vector bundle associated to this principal bundle and a linear action of G on a vector space V. We can use the connection to differentiate sections of W, producing one-forms with values in W. This covariant differentiation is a linear operator

$$\nabla_A \colon \Omega^0(B;W) \to \Omega^1(B;W).$$

To define this operator, suppose that σ is a section of $W \to B$. Near a point $x \in B$ we write

$$\sigma(b) = [p(b), v(b)]$$

where $p(b)$ and $v(b)$ are smooth functions of b with values in P and V respectively and with $p(b)$ being a local section of $P \to B$. Then the value $\nabla_A(\sigma)(b)$ evaluated on a tangent vector $\tau_b \in TB_b$ is equal to

$$[p(b), \omega_b(De_b(\tau_b))(v(b)) + Dv_b(\tau_b)]. \tag{1}$$

Here $De_b(\tau_b)$ is an element of TP_e, and the value of ω_b on this tangent vector is an element of $\mathcal{G}$. The differential at the identity of the linear representation $G \to Aut(V)$ is a Lie algebra map $\mathcal{G} \to End(V)$. Thus, we can evaluate $\omega_b(De_b(\tau_b))$ on $v(b)$ to obtain an element in V. A straightforward computation (Problem 20) shows that the quantity in Expression (1) is well-defined independent of the choice of $p(b)$ and $v(b)$ representing the section σ. Notice that if we choose $p(b)$ so that it is horizontal in the τ_b direction (as we always can) then Expression (1) simplifies to give

$$\nabla_A(\sigma)(\tau_b) = [p(b), \partial v/\partial \tau_b]$$

when $p(b)$ is horizontal in the τ_b direction. It is easy to see (Problem 21) that covariant differentiation satisfies the following differentiation law:

$$\nabla_A(f\sigma) = f\nabla_A(\sigma) + df \otimes \sigma \tag{2}$$

for any section σ of W and any real-valued function f on B. If V has a positive definite inner product $<,>$ and if G is contained in the orthogonal group $O(V)$ then in addition we have

$$d(<\sigma_1, \sigma_2>) = <\nabla_A(\sigma_1), \sigma_2> + <\sigma_1, \nabla_A(\sigma_2)>. \tag{3}$$

Given a principal G-bundle $P \to B$ with G a linear group, $G \subset Aut(V)$ a connection A on P is completely determined by the induced covariant derivative ∇_A on the associated vector bundle $P \times_G V$. Let $\tau_e \in TP_e$ be a non-vertical tangent vector. Choose any local section σ of P through e which has τ_e as a tangent vector. For $v \in V$ we consider

$$\nabla_A([\sigma, v])(D\pi_*(\tau_e)).$$

It can be written uniquely in the form $[\sigma(b), w]$. The assignment $v \mapsto w$ determines a linear endomorphism of V which in fact lies in $\mathcal{G} \subset End(V)$. This element is the value of $\omega_A(\tau_e)$. Conversely, one can use this equality in reverse to put conditions on a covariant differentiation on W which imply that it is induced from a connection on P (see Problem 22).

2.11. Problems

1. Suppose that $P \times G \to P$ is a smooth, free action of a compact Lie group. Show that this action gives rise to a smooth principal bundle. Give conditions on the action so that the conclusion remains true for non-compact Lie groups.
2. Let $P \times G \to P$ be a free action and let $\pi\colon P \to B$ be a G-invariant map that induces a homeomorphism between the quotient of the action and B. Show that this data determines a principal bundle if and only if there are local sections; i.e., for each point $b \in B$ there is a neighborhood $U \subset B$ of b and a continuous map $\sigma\colon U \to P$ with $\pi \circ \sigma$ equal to the inclusion of U into B.
3. Show that a principal G-bundle $P \to B$ is isomorphic to the product bundle $B \times G$ if and only if it has a cross section $\sigma\colon B \to P$.
4. State and prove the smooth analogue to problem 3.
5. Suppose that $\pi\colon P \to B$ is a topological principal bundle in which P, B and the action of G on P are smooth, and π is a smooth submersion. Show that this is a smooth principal bundle; i.e., show that the local trivializations can be chosen to be smooth.
6. Show that the formulas given for the action of $GL(n, \mathbf{R})$ on the space of frames E for the tangent bundle of a smooth manifold is indeed a right action and makes E a principal $GL(n, \mathbf{R})$ bundle.
7. Generalize the construction of the frame bundle of the tangent bundle of a manifold to produce a $GL(n, \mathbf{R})$ principal bundle P associated to any n-dimensional vector bundle V. Show that V is isomorphic to $P \times_{GL(n,\mathbf{R})} \mathbf{R}^n$. Show that this sets up a natural bijection between the set of isomorphism classes of real n-plane bundles over B and the set of isomorphism classes of principal $GL(n, \mathbf{R})$-bundles over B. State and prove analogous results for oriented bundles and complex bundles.
8. Define the grassmannian manifold of k-planes in n-space and define the tautological $O(k)$-bundle $\xi(n, k)$ over this space by giving an open cover over which the bundle is trivial and giving the transition functions for the bundle with respect to this cover. Define the grassmannian manifold of oriented k-planes in n-space and the tautological $SO(k)$-bundle over it. Give the complex and quaternion analogues of these bundles.
9. Show that the projection mapping for the tautological S^1-bundle over $\mathbf{C}P^1$ is isomorphic to the Hopf map $S^3 \to S^2$. Describe in similar terms the tautological unit quaternion bundle over $\mathbf{H}P^1$.

10. State and prove the homotopy covering property for locally trivial fiber bundles. Using this, state and prove the homotopy long exact sequence associated to a principal bundle.
11. Show that every (real) vector bundle has a positive definite inner product. Given such an inner product define the principal $O(n)$-bundle associated to the vector bundle. State and prove the analogous result for complex vector bundles.
12. Define principal bundles associated with the tangent bundle of a symplectic manifold and of a spin manifold with structure groups $Sympl(n)$ and $Spin(n)$ respectively.
13. Show that the cocycle condition for transition functions implies that the relation given in defining a principal bundle from the local trivializations is indeed an equivalence relation.
14. Show that cocycles which differ by a coboundary define isomorphic principal bundles.
15. Show that if $F\colon Q \to P$ is a G-equivariant map between the total spaces of principal G-bundles over A and B respectively, then F induces a map $f\colon A \to B$ and an isomorphism of principal G-bundles $Q \to f^*P$.
16. Suppose that $\pi\colon P' \to B$ is a principal H-bundle and $H \subset G$ is a sub-Lie group. Show that $P' \times_H G \to B$ is naturally a principal G-bundle. A reduction of a G-bundle $P \to B$ to an H-bundle is a pair: an H-bundle $P' \to B$ and an isomorphism of G-bundles $P' \times_H G \to P$. Show that a principal G-bundle reduces to the subgroup $H = \{1\}$ if and only if the G-bundle is trivial.
17. Show that reducing a $GL(n, \mathbf{R})$-bundle to an $O(n)$-bundle is equivalent to choosing a positive definite inner product on the associated vector bundle.
18. Show that any vector bundle over a compact base space embeds as a subvector bundle of a trivial vector bundle. Using this show that any n-dimensional bundle over a compact base space is isomorphic to one induced from the tautological bundle over the grassmannian of n-planes in $\mathbf{R}^k$ for some $k >> 0$. Show the analogous result for smooth bundles.
19. Show that pulling back via the projection π induces an isomorphism between $\Omega^1(B; \mathrm{ad}P)$ and the subspace of $\Omega^1(P; \mathcal{G})$ consisting of all one-forms which transform under the adjoint representation when transported by $g \in G$ and which vanish on vertical tangent vectors.
20. Show that the expression given in Equation 1 gives a well-defined value independent of the choices of $p(b)$ and $v(b)$ representing σ.
21. Show that Equation 2 holds.
22. Show that a covariant differentiation on a vector bundle with a positive definite inner product is induced from a connection on the underlying principal $O(V)$-bundle if and only if Equation (3) holds. Give corresponding statements for special orthogonal bundles, unitary bundles, and special unitary bundles.
23. Suppose ∇ is a connection on the tangent bundle of a smooth manifold M. We define the *torsion*, T of ∇ by

$$T(X, Y) = \nabla_x(Y) - \nabla_y(X) - [X, Y]$$

for all vector fields X, Y on M. The connection ∇ is said to be *torsion-free* if $T(X, Y) = 0$ for all X, Y.

Show that the tangent bundle of a remannian manifold has a unique connection which is torsion-free and orthogonal, i.e. compatible with the metric. The connection is called the *Levi-Cività connection.*

24. Compute the cohomology ring with rational coefficients of $Gr^{+}_{\mathbf{R}}(n, k)$ the grassmannian of oriented n-planes in $\mathbf{R}^{n+k}$ for any $k \leq \infty$. Define the rational Pontrjagin classes of an oriented vector bundle.
25. Compute the cohomology ring with integer coefficients of $Gr_{\mathbf{C}}(n, k)$ of complex n planes in $\mathbf{C}^{n+k}$ for any $k \leq \infty$. Define the Chern classes of a complex vector bundle.
26. Let $\pi \colon P \to B$ be a principal G-bundle. Show that $\pi^* P \to P$ is a trivial principal G-bundle.
27. Complete the proof of Lemma 2.8.1.

LECTURE 3
Curvature and Characteristic Classes

3.1. The curvature of a connection as an obstruction to integrating the horizontal distribution

In our first description, the curvature arises as the obstruction to integrating the horizontal distribution of a connection over two-dimensional submanifolds of the base.

Let $P \to B$ be a smooth principal G-bundle and let $\mathrm{ad}P$ be the vector bundle associated to P and the adjoint action of G on its Lie algebra $\mathcal{G}$. Suppose that A is a connection on P, which we view for the moment as a horizontal distribution $\mathcal{H} \subset TP$. As we have seen, we can integrate $\mathcal{H}$ along paths in B to give us a lifting of paths from B to P. If we try to perform the same construction over higher dimensional subspaces of B then it is not always possible to lift – there is an obstruction which is the curvature of the connection. Let us fix a point $b \in B$ and two linearly independent tangent vectors τ_1, τ_2 at b. Consider a local coordinate system $(x_1, \ldots, x_k)$ centered at a point $b \in B$ with the property that

$$(\partial/\partial x_i)|_0 = \tau_i \quad \text{for} \quad i = 1, 2.$$

We consider a rectangle $[0, \epsilon] \times [0, \epsilon]$ in the (x_1, x_2)-subspace. We lift the four sides of this rectangle in counterclockwise fashion beginning with the side on the x_1-axis. We do this so that the initial point lifts to a point $p \in P$ and so that each side begins where the previous side ends. There is no guarantee that the end of the last side will be equal to p, but it will be of the form $p \cdot g$ for some unique $g = g(\epsilon) \in G$.

If ϵ is sufficiently close to zero, then $g(\epsilon)$ will be close to the identity in G, and hence we can form $\log(g(\epsilon)) \in \mathcal{G}$. We consider the element

$$K_A(\epsilon) = -\frac{\log(g(\epsilon))}{\epsilon^2}.$$

Lemma 3.1.1. *The element in g given by*

$$K_A(p, \tau_1, \tau_2) = \lim_{\epsilon \mapsto 0} K_A(\epsilon)$$

depends only on p, τ_1, τ_2. Furthermore, the point

$$[p, K_A(e, \tau_1, \tau_2)] \in \mathrm{ad}P$$

depends only on τ_1, τ_2*, and is bilinear and skew-symmetric in these variables. It is given by evaluating a two-form on* B *with values in* ad$P$*, denoted* F_A*, on* (τ_1, τ_2)*. This two-form* F_A *is called the* curvature *of* A*.*

Remark 3.1.2. What is the explanation for the minus sign in this formula? This is what is used, but why is a mystery to the author.

Proof. Work over the square $S = [0, \epsilon] \times [0, \epsilon]$ and choose a trivialization of $P|_S$ which is parallel with respect to the connection A in the x_1-direction over $[0, \epsilon] \times \{0\}$ and which is parallel with respect to the connection in the x_2-direction at every point of the square. Restricting the connection one-form on P to the square $S \times \{1\}$ in this trivialization gives us a one-form

$$A_1 dx_1 + A_2 dx_2$$

where the A_i are smooth functions of (x_1, x_2) with values in $\mathcal{G}$. The conditions we have imposed on the square imply that

$$A_2 = 0, \text{ and } A_1|_{[0,\epsilon]\times\{0\}} = 0.$$

One easily computes (See Problem 1) the limit in the statement of the lemma at the origin of S and the identity of G to be equal to $\partial A_1 / \partial x_2(0, 0)$. All the claimed properties follow from this computation. □

We can reformulate the limit of what happens around small squares in terms of the Lie bracket of vector fields. Suppose that χ_1, χ_2 are vector fields on B. These lift uniquely to horizontal vector fields $\widetilde{\chi}_1, \widetilde{\chi}_2$ on P. The bracket $\widetilde{\chi} = [\widetilde{\chi}_1, \widetilde{\chi}_2]$ is a vector field on P. We use the horizontal subspaces to project this vector field to a vertical vector field $\widetilde{\chi}^v$ on P.

Lemma 3.1.3. *The vector field* $\widetilde{\chi}^v$ *is a* G*-invariant vector field on* P*. As such it is equivalent to a section* $\sigma(\chi_1, \chi_2)$ *of* ad$P$*.*

Proof. Since the family of horizontal spaces is G-invariant, it follows that $\widetilde{\chi}_i$ is also G-invariant for $i = 1, 2$. Thus, $\widetilde{\chi}$ is also G-invariant. Again using the G-invariance of the horizontal distribution, we see that $\widetilde{\chi}^v$ is also G-invariant. This determines a section of ad$P \to B$ by the formula

$$[p, (DR_p)^{-1}(\widetilde{\chi}^v_p)].$$

We need to see that this expression is a well-defined section of ad$P$. This follows from the equations

$$[pg, (DR_{pg})^{-1}(\widetilde{\chi}^v_{pg})] = [pg, (DR^{-1}_{pg})(\widetilde{\chi}^v_p \cdot g)] = [pg, g^{-1} \cdot (DR_p)^{-1}(\widetilde{\chi}^v_p) \cdot g].$$

□

Now we are ready to connect this with the curvature.

Lemma 3.1.4. *Keeping the notation from above, we have*

$$-\sigma(\chi_1, \chi_2) = F_A(\chi_1, \chi_2).$$

Proof. The first thing to notice is that $\sigma(\chi_1, \chi_2)$ is bilinear over the smooth functions acting on the vector fields. We have $\widetilde{(f\chi)} = f\widetilde{\chi}$. Hence,

$$[\widetilde{(f_1\chi_1)}, \widetilde{(f_2\chi_2)}] = f_1 f_2 [\widetilde{\chi}_1, \widetilde{\chi}_2] + h$$

where h is a horizontal vector field (see Problem 13). Projecting this equation gives the required bilinearity. From this it follows that the value of $\widetilde{\chi}^v(p)$ depends only on the values at p of χ_1, χ_2.

We use the same local coordinates $(x_1, \ldots, x_n)$ and local trivialization that we used in proving that the limit exists. By the fact just established, we can assume that $\chi_1 = \partial/\partial x_i$. Restricted to the square $S \times \{1\}$ as before we have $\widetilde{\chi}_2 = \partial/\partial x_2$ and $\widetilde{\chi}_1|_{[0,\epsilon]\times\{0\}} = 0$. Thus, the value of the bracket at $(0,0)$ is equal to $-\partial A_1/\partial x_2$ at the origin. Proves the lemma. □

Corollary 3.1.5. *The value of the curvature F_A on two vector fields χ_1, χ_2 on B is simply the negative of the vertical projection (viewed in the usual way as an element of $\mathcal{G}$) of the bracket of the horizontal lifts of the vector fields. In particular according to the Frobenius Theorem, the distribution $\mathcal{H}_A$ is integrable if and only if $F_A = 0$. In this case the connection A is said to be* flat.

3.2. Interpretation of the curvature in terms of the connection-one form

In another direction, let us relate the curvature F_A to the connection one-form $\omega_A \in \Omega^1(P;\mathcal{G})$. First, we need a lemma which extends Lemma 3.1.3.

Lemma 3.2.1. *Pullback via the projection mapping π induces an isomorphism between $\Omega^k(B;\mathrm{ad}P)$ and the subspace of $\Omega^k(P;\mathcal{G})$ consisting of all forms μ which satisfy:*

- $\mu(\tau_1 \cdot g, \ldots, \tau_k \cdot g) = g^{-1} \cdot \mu(\tau_1, \ldots, \tau_k) \cdot g$, *and*
- $\mu(\tau_1, \ldots, \tau_k) = 0$ *if τ_1 is a vertical vector.*

Proof. See Problem 2. □

Let $\eta \in \Omega^1(B;\mathrm{ad}P)$. We wish to define a two-form on B with values in $\mathrm{ad}P$ which is denoted both as $\eta \wedge \eta$ and as $(1/2)[\eta,\eta]$. First note that the Lie bracket on $\mathcal{G}$ induces a well-defined skew-symmetric bundle map

$$\frac{1}{2}[\ ,\ ] : \mathrm{ad}P \otimes \mathrm{ad}P \to \mathrm{ad}P.$$

The value of $\eta \wedge \eta$ on a pair of tangent vectors τ_1, τ_2 is given by

$$[\eta(\tau_1), \eta(\tau_2)].$$

Thus, in forming this product we are using both the wedge product of forms and the Lie bracket in $\mathcal{G}$ (or $\mathrm{ad}P$).

Lemma 3.2.2. *The two-form π^*F_A on P with values in $\mathcal{G}$ is equal to*

$$d\omega_A + \omega_A \wedge \omega_A.$$

Proof. Let us consider $\pi^*F_A \in \Omega^2(P;\mathcal{G})$. We have already seen that π^*F_A evaluates on a pair of horizontal vector fields to give minus the vertical projection of their bracket. Also, π^*F_A vanishes on a pair of vector if one of them is vertical. It is clear that these properties uniquely determine π^*F_A among all two-forms on P with values in $\mathcal{G}$. Thus, our task is to show that $d\omega_A + \omega_A \wedge \omega_A$ also has these properties.

First let us consider horizontal vector fields χ_1, χ_2. Since $\omega(\chi_i) = 0$, we see that $\omega_A \wedge \omega_A(\chi_1, \chi_2) = 0$ and $d\omega(\chi_1, \chi_2) = -\omega([\chi_1, \chi_2])$. This proves the result for a pair of horizontal vector fields.

Now suppose that τ_1 is a vertical at $p \in P$ and τ_2 is a horizontal vector at p. Take a horizontal curve J through p with τ_2 as tangent vector. We can restrict the principal bundle over the curve in order to compute

$$(d\omega + \omega_A \wedge \omega_A)(\tau_1, \tau_2).$$

Parallel transport induces a trivialization $P|_J = J \times G$ sending the connection to the product connection. Thus, in this trivialization ω becomes the pullback from the G factor of the Maurer-Cartan form. Likewise, $d\omega + \omega_A \wedge \omega_A$ is pulled back from the G factor and hence vanishes on the pair (τ_1, τ_2) since τ_2 is tangent to the J factor.

Lastly, suppose that τ_1, τ_2 are both vertical. In this case, we can restrict to a single fiber and identify this fiber with a copy of G. The connection one-form becomes the Maurer-Cartan form on the group. Thus, the result follows from the fact that $d\omega_{MC} + \omega_{MC} \wedge \omega_{MC} = 0$. This is left as an exercise (Problem 10.). □

We can view this computation as a computation of the curvature in terms of a local trivialization for P over an open subset $U \subset B$. Suppose that we have a trivialization $P|_U \cong U \times G$. We pull the curvature one-form ω_A back to $U \times G$ and restrict to $U \times \{1\}$. In this way we obtain a one form $\alpha \in \Omega^1(U; \mathcal{G})$. This trivialization induces one of $\text{ad}P|_U$. Under this trivialization the curvature $F_A|_U$ becomes a two-form on U with values in $\mathcal{G}$.

Lemma 3.2.3. *With the above conventions we have*

$$d\alpha + \alpha \wedge \alpha = F_A|_U.$$

Proof. This is easily established by restricted from $\pi^* F_A$ on P and using the previous result. □

3.3. The relationship of curvature and covariant differentiation

Let us give a description of F_A in terms of covariant differentiation. Suppose that we have a representation ρ of G on a vector space V. Let $W \to B$ be the associated vector bundle $P \times_G V \to B$. Recall that the connection A gives rise to a covariant derivative

$$\nabla_A \colon \Omega^0(B; W) \to \Omega^1(B; W).$$

This derivation extends uniquely to a linear operator

$$\nabla_A \colon \Omega^k(B; W) \to \Omega^{k+1}(B; W)$$

satisfying

$$\nabla_A(\eta \wedge \lambda) = d\eta \wedge \lambda + (-1)^r \eta \wedge \nabla_A(\lambda)$$

for any $\eta \in \Omega^r(B; \mathbf{R})$ and $\lambda \in \Omega^{k-r}(B; W)$.

Lemma 3.3.1. *The linear operator*

$$\nabla_A \circ \nabla_A \colon \Omega^0(B; W) \to \Omega^2(B; W)$$

is given by multiplying by a two-form with values in End$E$. *In fact,*

$$\nabla_A \circ \nabla_A(\sigma) = \text{ad}\rho(F_A)(\sigma)$$

for any $\sigma \in \Omega^0(B; W)$.

Proof. Let σ be a section of W and f a smooth function on B. We expand

$$\begin{aligned}\nabla_A \circ \nabla_A(\sigma \otimes f) &= \nabla_A(\nabla_A(\sigma) \otimes f + \sigma \wedge df) \\ &= \nabla_A \circ \nabla_A(\sigma) \otimes f - \nabla_A(\sigma) \wedge df + \nabla_A(\sigma) \wedge df - \sigma \wedge d^2 f \\ &= \nabla_A \circ \nabla_A(\sigma) \otimes f\end{aligned}$$

This shows that $\nabla_A \circ \nabla_A$ is linear over the functions on B, and hence is given by evaluating a 2-form on B with values in the endomorphism bundle $End(V) \to B$. To calculate this two-form, we lift to the principal bundle. The pullback $\pi^* W$ is canonically identified with the trivial bundle $P \times V$ (see Problem 9). Let $\widetilde{\sigma} \in \Omega^0(P; V)$ be the pullback of σ. Let ω be the connection one-form. We have that the pullback of $\nabla_A \circ \nabla_A(\sigma)$ is equal to

$$(d + \omega) \circ (d + \omega)(\widetilde{\sigma}) = d(d\widetilde{\sigma} + \omega(\widetilde{\sigma})) + \omega(d\widetilde{\sigma} + \omega(\widetilde{\sigma})).$$

The latter term is equal to

$$d\omega(\widetilde{\sigma}) + \omega(\omega(\widetilde{\sigma})).$$

According to Problem 8 this completes the proof. □

3.4. Characteristic classes

We can use the curvature to define cohomology classes in B which measures the non-triviality of the bundle. These are called *characteristic classes*. The first result we need in order to define characteristic classes from the curvature is the so-called Bianchi identity.

Lemma 3.4.1. *(Bianchi Identity)* $\nabla_A F_A = 0$.

Proof. Work with $F_A = d\omega_A + (1/2)[\omega, \omega]$ on P and use the fact that $\nabla_A F_A = dF_A + [\omega_A, F_A]$. Then we have

$$\begin{aligned}\nabla_A F_A &= d(d\omega_A + (1/2)[\omega_A, \omega_A]) + [\omega_A, d\omega_A + (1/2)[\omega_A, \omega_A]] \\ &= (1/2)[d\omega_A, \omega_A] - (1/2)[\omega_A, d\omega_A] + [\omega_A, d\omega_A] + (1/2)[\omega_A, [\omega_A, \omega_A]] \\ &= (1/2)[\omega_A, d\omega_A] + (1/2)[d\omega_A, \omega_A] + (1/2)[\omega_A, [\omega_A, \omega_A]]\end{aligned}$$

The sum of the first two terms vanishes by the graded skew-symmetry of the Lie bracket on forms:

$$[A, B] = -(-1)^{deg(A) \cdot deg(B)}[B, A].$$

The last term vanishes by the graded Jacobi identity. □

Now suppose that

$$\varphi\colon \underbrace{\mathcal{G} \otimes \cdots \otimes \mathcal{G}}_{k \text{ times}} \to \mathbf{R}$$

is a linear map which is symmetric and invariant under the simultaneous adjoint action of G on $\mathcal{G}$; i.e.,

$$\varphi(F_1, \ldots, F_k) = \varphi(g^{-1}F_1 g, \ldots, g^{-1}F_k g).$$

Then we can form

$$\varphi(F_A, \ldots, F_A) \in \Omega^{2k}(B; \mathbf{R}).$$

Lemma 3.4.2. *The form $\varphi(F_A, \ldots, F_A)$ is closed. If we choose another connection A' for P then the difference*

$$\varphi(F_{A'}, \ldots, F_{A'}) - \varphi(F_A, \ldots, F_A)$$

is exact.

Proof. To prove that this form is closed is an exercise (Problem 5). To prove that the forms associated to connections A and A' differ by an exact form we construct a connection $\overline{A}$ for the bundle $P \times I \to B \times I$ which agrees with A on $P \times \{0\}$ and agrees with A' over $P \times \{1\}$ (see Problem 6). The form $\overline{\mu} = \varphi(F_{\overline{A}}, \ldots, F_{\overline{A}})$ is a closed form on $B \times I$ whose restrictions to the two ends are $\varphi(F_A, \ldots, F_A)$ and $\varphi(F_{A'}, \ldots, F_{A'})$. Integrating $\overline{\mu}$ in the I direction gives a form on B of one lower degree whose exterior differential is $\varphi(F_{A'}, \ldots, F_{A'}) - \varphi(F_A, \ldots, F_A)$. □

For the special orthogonal group $SO(n)$, a basis for the invariant polynomials on the Lie algebra is given by the even coefficients of the characteristic polynomial (by skew-symmetry the odd coefficients vanish) together with the Pfaffian if n is even. Invoking the above construction, we get one characteristic class in each degree $4i$, and if $n = 2k$ we also get one characteristic class in degree $2k$. If we normalize properly then these classes are respectively the ith Pontrjagin class and the Euler class.

There is a similar result for complex-valued symmetric, multilinear functions on the Lie algebra. Applying this to the unitary group we see that a basis for the complex-valued invariant polynomials is given by the coefficients of the characteristic polynomial. Thus, in this case we have one characteristic class in each degree $2i$. Correctly normalized these are the Chern classes.

Notice that all the classes are integral classes. Of course, this is not obvious from the construction given here since we are working with differential forms (either real or complex) and these represent real or complex cohomology. But it is a miracle of classifying spaces that the classes defined this way and suitably normalized are in fact the images of integral classes.

3.5. The holonomy of a connection

Let A be a connection on a smooth principal G-bundle $P \to B$. Let $p \in P$. The *holonomy* of A at p is a function from the loop space $\Omega(B, \pi(p))$ to the group G. Its value on the loop λ based at $x = \pi(p)$ is determined by taking the horizontal lift $\widetilde{\lambda}$ beginning at p and considering the other end point $\widetilde{\lambda}(1)$. This is a point in the fiber over x, and as such it can be written uniquely as $p \cdot h_{A,p}(\lambda)$ for some $h_{A,p}(\lambda) \in G$. This defines $h_{A,p}$, the holonomy of A at p.

Lemma 3.5.1. *The holonomy $h_{A,p}\colon \Omega(B, x) \to G$ is an anti-homomorphism in the sense that for any loops $\lambda, \mu \in \Omega(B, x)$ we have*

$$h_{A,p}(\lambda * \mu) = h_{A,p}(\mu) \cdot h_{A,p}(\lambda)$$

and $h_{A,p}(\lambda^{-1}) = h_{A,p}(\lambda)^{-1}$.

Proof. Since horizontal lifting is G-equivariant, the horizontal lift of μ beginning at $p \cdot h_{A,p}(\lambda)$ ends at $p \cdot (h_{A,p}(\mu) \cdot h_{A,p}(\lambda)$. The inverse property is obvious. □

Corollary 3.5.2. *The image of the holonomy representation is a subgroup $\mathrm{Hol}_{A,p} \subset G$.*

This computation also has the following consequence:

Lemma 3.5.3. *The holonomy map satisfies:*

$$h_{A,pg} = g^{-1} \cdot h_{A,p} \cdot g.$$

3.6. Problems

1. Do the computation omitted in the proof of Lemma 3.1.1.
2. Prove Lemma 3.2.1
3. Let V_1 and V_2 be vector bundles over a base B, and let ∇_1 and ∇_2 be covariant derivatives over V_1 and V_2 respectively. (a) Show that there is a (unique) naturally defined covariant derivative ∇_{hom} on $Hom(V_1, V_2)$ which satisfies Leibnitz's rule for the pairing
$$<,>: \Omega^0(B, V_1) \otimes \Omega^0(B, Hom(V_1, V_2)) \to \Omega^0(B, V_2);$$
i.e.
$$\nabla_2 < \varphi, s > = < \nabla_{hom}\varphi, s > + < \varphi, \nabla_1(s) > .$$
Moreover, show that there is an induced pairing
$$<,>: \Omega^i(B, V_1) \otimes \Omega^j(B, Hom(V_1, V_2)) \to \Omega^{i+j}(B; V_2),$$
and that for this pairing, the covariant derivative satisfies the graded Leibnitz rule
$$\nabla_2 < \omega, \eta > = < \nabla_{hom}(\omega), \eta > + (-1)^{\deg \omega} < \omega, \nabla_1 \eta > .$$
(b) Similarly, show that ∇_1 and ∇_2 induce a connection $\nabla_\otimes$ on $V_1 \otimes V_2$ which satisfies the graded Leibnitz rule induced by the pairing
$$\Omega^0(V_1) \otimes \Omega^0(V_2) \to \Omega^0(V_1 \otimes V_2)$$
4. Let P be a principal G-bundle and V be a G-vector space. If $v \in V$ is fixed by the G-action, show that there is a naturally induced section (also denoted v) of $W = P \times_G V$. Show that for any covariant derivative ∇_A on W induced from a connection A on P, $\nabla_A v = 0$.
5. Apply the previous two problems and the Bianchi identity to prove that if φ is a G-invariant, symmetric, linear map
$$\varphi\colon \underbrace{\mathcal{G} \otimes \cdots \otimes \mathcal{G}}_{k \text{ times}} \to \mathbf{R},$$
then the induced $2k$-form $\varphi(F_A, \ldots, F_A)$ is closed.
6. Let A_0 and A_1 be connections on a principal bundle $P \to B$. Show that there is a connection A on $P \times I \to B \times I$ with the property that $A|_{P \times \{i\}} = A_i$ for $i = 0, 1$.
7. Let V be a representation of G, and let $W = P \times_G V$. Show that there is a naturally induced map
$$\mathrm{ad}P \to End(W),$$
hence a pairing
$$<,>: \Omega^i(B; W) \otimes \Omega^j(B; \mathrm{ad}P) \to \Omega^{i+j}(B; W).$$
8. For $\omega \in \Omega^2(End(W))$, $s \in \Omega^0(W)$, what is meant by $(1/2)[\omega, \omega](s)$, by $\omega(\omega(s))$, and by $\omega \wedge \omega(s)$? Show that all three quantities are equal.

9. Let $P \to B$ be a principal G bundle. and let $W = P \times_G V$ be an associated vector bundle. Show that $\pi^* W \to P$ is canonically identified with the trivial bundle $P \times V \to P$. (Compare with problem 26 of Chapter II.)
10. Show that the Maurer-Cartan form on a Lie group satisfies:

$$d\omega_{MC} + \omega_{MC} \wedge \omega_{MC} = 0.$$

11. Show that a flat connection is induced from the product connection on the trivial bundle over the universal covering $\widetilde{B}$ of B by a representation of $\pi_1(B) \to G$. This means that P is isomorphic to

$$\widetilde{B} \times_{\pi_1(B)} G$$

where $\widetilde{B}$ is the universal covering of B; $\pi_1(B)$ acts on $\widetilde{B}$ as covering transformations and acts on G via left multiplication by some representation of $\pi_1(B) \to G$. Under this isomorphism the connection corresponds to the tangent spaces of the images of $\widetilde{B} \times \{pt\}$ in the total space of the principal bundle. In particular, show that if B is simply connected then any flat connection is isomorphic to the trivial connection.
12. Let $W \to B$ be a vector bundle. Suppose that we have a skew-symmetric, multilinear function of vector fields on B:

$$\sigma\colon \underbrace{\mathrm{Vect}(B) \otimes_B \cdots \otimes_B \mathrm{Vect}(B)}_{k \text{ times}} \to W.$$

Show that σ is given by evaluating a k-form on B with values in W if and only if σ is linear over multiplication by functions in each variable.
13. Show that if $\widetilde{\chi}_1, \widetilde{\chi}_2$ are horizontal vector fields, then

$$[\widetilde{(f_1\chi_1)}, \widetilde{(f_2\chi_2)}] = f_1 f_2 [\widetilde{\chi}_1, \widetilde{\chi}_2] + h$$

where h is a horizontal vector field.
14. Let $A \in u(n)$, the Lie algebra of $U(n)$. The expression

$$\mathrm{Det}(\frac{1}{2\pi i}(\mathrm{Id} \cdot x + A))$$

is a polynomial in the variable x with coefficients which are polynomials in the entries of A. Of course, these polynomials are invariant under the adjoint action since the determinant is. The coefficient of x^k is homogeneous of degree $(n-k)$ on $u(n)$. It is the homogeneous function which determines the $(n-k)$ Chern class c_{n-k} for principal $U(n)$-bundles. Show that under the inclusion $U(n) \subset U(n+1)$ the Chern classes c_k correspond. Evaluate these classes on the tautological bundle over the complex grassmannian $Gr_{\mathbf{C}}(n, \infty)$
15. Prove that if $V \to B$ is an n-dimensional complex bundle associated to a hermitian principal bundle P, then the class $c_n(P)$ is equal to the Euler class of V.
16. Let $P \to B$ be a smooth principal $U(1) = S^1$-bundle. We identify the Lie algebra of $U(1)$ with the totally imaginary complex numbers. Show that the Maurer-Cartan form is $id\theta$ on S^1 so that

$$\int_{S^1} \omega_{\mathrm{MC}} = 2\pi i.$$

Let ω_A be a connection one-form on P. Then show that for any point $b \in B$ we have

$$\frac{1}{2\pi i}\int_{P_b} \omega_A = 1,$$

where $F_b \subset P$ is the fiber over b. Show that the two-form $d(\omega_A/2\pi i)$ on P is induced from a closed two-form on B and that this two-form represents the first Chern class of the line bundle.

17. Give the normalization of the Pontrjagin classes.
18. Compute the Pontrjagin classes of the universal bundle $\xi_n^{\mathbf{R}} \to Gr_{\mathbf{R}}(n,\infty)$.
19. Show that $p_n(\xi) = (-1)^n c_{2n}(\xi \otimes_{\mathbf{R}} \mathbf{C})$. Show that $2c_{2n+1}(\xi \otimes_{\mathbf{R}} \mathbf{C}) = 0$.
20. Prove the product formula for the Chern classes and Stiefel-Whitney classes
$$c_*(\xi \oplus \eta) = c_*(\xi) \cdot c_*(\eta).$$
Prove the corresponding statement for the Pontrjagin classes modulo 2-torsion.
21. Classify up to isomorphism all principal $SU(2)$ and $SO(3)$-bundles over a finite 4-dimensional simplicial complex X.
22. Show that a connection A for P is flat if and only if for every $p \in P$ and $x = \pi(p) \in B$ the holonomy representation:
$$h_{A,p} \colon \Omega(B,x) \to G$$
factors through the map
$$\Omega(B,x) \to \pi_1(B,x)$$
induced by taking homotopy classes.
23. Let $H \subset G$ be a closed subgroup. Show that a connection A on a principal G bundle $P \to B$ reduces to a connection on a principal H-bundle Q if and only if for every $p \in P$ the image of the holonomy representation $h_{A,p}$ is contained in a conjugate of H.

LECTURE 4
The Space of Connections

The group of bundle isomorphisms, its action on the space of connections, and the quotient space of eigenvalue classes of connections

4.1. The group of bundle isomorphisms of P

Let $\pi\colon P \to B$ be a smooth principal G-bundle. An *automorphism* of P is a diffeomorphism $\varphi\colon P \to P$ which is G-equivariant and which covers the identity on the base in the sense that $\pi \circ \varphi = \pi$. Clearly, the set of automorphisms of P forms a group under composition. This group acts on the left on P commuting with the right action of G. We denote this group by $\mathrm{Aut}(P)$. It is called the *group of gauge transformations*. The source of the terminology is that in physics (where these objects were introduced about 40 years ago) a frame was considered as a gauge by which measurements could be made. Thus, one could change the gauge. The group of all changes of gauge is the group we are introducing here. In common parlance, this group is often misnamed the *gauge group*. In fact, the gauge group is the original Lie group G.

We have the following non-linear version of the vector bundle $\mathrm{ad}P$.

Lemma 4.1.1. *Let* $\mathrm{Ad}P$ *be the locally trivial fiber bundle*

$$P \times_G G \to B$$

where G acts on itself by conjugation: $g \cdot h = ghg^{-1}$. This is a locally trivial fiber bundle of groups; i.e., the fibers are groups and the transition functions map the intersections of the open subsets of B to the space of group isomorphisms of the fiber. Consequently, the space of C^∞-sections of this fiber bundle inherits a continuous group multiplication from the fiber-wise multiplications.

The proof is straightforward.

The point of introducing this non-linear bundle of groups is that, as the next lemma shows, its space of sections is naturally identified with the group of gauge transformations.

Lemma 4.1.2. *The group of C^∞-sections of* AdP *under fiber-wise multiplication is naturally isomorphic to the group* Aut(P).

Proof. Let $\varphi\colon P \to P$ be an isomorphism. Since p and $\varphi(p)$ lie in the same fiber of P for each $p \in P$, there is a unique $\psi\colon P \to G$ such that for all $p \in P$ we have

$$\varphi(p) = p \cdot \psi(p).$$

Since φ is a diffeomorphism, it follows that ψ is C^∞. Since $\varphi(pg) = \varphi(p) \cdot g$ it follows that

$$\psi(pg) = g^{-1}\psi(p)g. \tag{4}$$

Conversely, given any C^∞-function ψ satisfying Equation 4.1, one can define an automorphism of P by the formula

$$\varphi(p) = p \cdot \psi(p).$$

This identifies the group Aut(P) with the group of all ψ satisfying 4.1. The multiplication on the ψ is induced from the multiplication on G. On the other hand, it is easy to see that the group of such ψ is exactly the group of C^∞-sections of AdP under fiber-wise multiplication. □

The group Aut(P) is not a classical Lie group since it is infinite dimensional. It is however and infinite dimensional Lie group. In order to say what we mean by this we need to introduce a little of the theory of infinite dimensional manifolds.

4.2. Infinite dimensional manifolds, a first version

There is a natural theory of infinite dimensional manifolds. In this theory the space of C^∞-sections of a locally trivial smooth fiber bundle is an infinite dimensional Frechet manifold with the C^∞-topology. Applying this theory then, one sees that Aut(P) is an infinite dimensional manifold. It is also clear that the multiplication and inverse operations in the group are smooth maps. Thus, in some sense, this group is an infinite dimensional Lie group. It is natural to think that its Lie algebra is the vector space of C^∞-sections of ad$P$. Indeed this is true, but one has to take some care in establishing it, because we are working in the context of infinite dimensional manifolds. It is clear that there is a fiberwise exponential map:

$$\exp\colon \Omega^0(B; \mathrm{ad}P) \to \mathrm{Aut}(P)$$

which assigns to any section

$$\sigma\colon B \to \mathrm{ad}P$$

the section

$$s\colon B \to \mathrm{Ad}P$$

where $s(x) = \exp(\sigma(x))$. To see that this equation gives a well-defined correspondence, one uses the fact that the map $\exp\colon \mathfrak{g} \to G$ satisfies

$$\exp(g^{-1}Ag) = g^{-1}\exp(A)g$$

for all $g \in G$ and $A \in \mathfrak{g}$.

We define the Lie bracket structure on the vector space of sections of ad$P$ by fiberwise bracket. All the Lie algebra axioms clearly hold since they hold on each fiber. It is also clear that the usual Baker-Campbell-Hausdorff formula relating fiberwise multiplication of sections of AdP and fiberwise Lie bracket for sections of ad$P$ holds. Lastly, the general theory of infinite dimensional manifolds (See for example [8]) shows that the exponential map identifies the space of sections of ad$P$ with the tangent space at the identity section of the space of sections of AdP. In particular, any section of AdP sufficiently close to the identity section is the image under exponentiation of a unique small section of ad$P$, and exponentiation induces a diffeomorphism between a neighborhood of 0 in the vector space of all sections of ad$P$ and a neighborhood of the identity in Aut(P).

It turns out that for serious analytic application, these C^∞ objects are not the right ones to work with — Frechet spaces are cumbersome. When we get down to doing real analytic work, making estimates and so on, we will work with different infinite versions of these manifolds which are Hilbert manifolds (modeled on Hilbert spaces). I have not yet introduced these objects here because they are more complicated to define, and at this point it is not clear that the extra complication is justified. But when the time is ripe we will work with other versions of these spaces.

4.3. The action of the group of gauge of transformations on the space of connections

Let $\sigma \in \mathrm{Aut}(P)$ be an element of the group of gauge transformations and let $\omega \in \Omega^1(P;\mathfrak{g})$ be a connection one-form. Then we have $\sigma^*\omega \in \Omega^1(P;\mathfrak{g})$.

Lemma 4.3.1. *The form $\sigma^*\omega$ is a connection one-form for P.*

Proof. First, since σ commutes with the G-action and since ω transforms via the adjoint action of G on $\mathfrak{g}$, the same is true for $\sigma^*\omega$. Second, if we identify a fiber of $P \to B$ with G by choosing a point e in the fiber and identifying g with $e \cdot g$, then restriction of σ to this fiber is identified with left multiplication by an element of G. The restriction of ω to this fiber is identified with the left invariant Maurer-Cartan form. Thus, the restrictions of $\sigma^*\omega$ and of ω to this fiber agree. This shows that the two conditions required for $\sigma^*\omega$ to be a connection one-form are satisfied. □

If we view a connection A as a horizontal distribution $\mathcal{H}_A$, then the

$$(\mathcal{H}_{\sigma^*(A)})_p = D\sigma_p^{-1}((\mathcal{H}_A)_{\sigma(p)}).$$

One sees easily that this distribution is horizontal and is G-invariant. In terms of covariant differentiation in associated vector bundles, we have

$$\nabla_{\sigma^*A}(s) = \sigma^{-1}\left(\nabla_A(\sigma(s))\right).$$

Identifying the space of connections on P, which we will henceforth denote by $\mathcal{A}(P)$, with the (affine) space of connection 1-forms in $\Omega^1(P;\mathfrak{g})$, the last lemma has defined a map

$$\mathcal{A}(P) \times \mathrm{Aut}(P) \to \mathcal{A}(P).$$

Lemma 4.3.2. *If $\sigma, \tau \in \mathrm{Aut}(P)$ and if A is a connection on P, then*

$$\tau^*(\sigma^*(A)) = (\sigma \circ \tau)^*(A).$$

That is to say, pulling back connections induces a right action

$$\mathcal{A}(P) \times \mathrm{Aut}(P) \to \mathcal{A}(P)$$

where $\mathcal{A}(P)$ denotes the space of connections on P.

Proof. Obvious. □

The action of $\mathrm{Aut}(P)$ on P induces is a natural left action of $\mathrm{Aut}(P)$ on any fiber bundle associated to P. It is given by $\sigma \cdot [p, v] = [\sigma(p), v]$ (see Problem 1). In particular, there is an action of this group on $\mathrm{ad}P$. This action induces a linear action ρ of $\mathrm{Aut}(P)$ on $\Omega^1(B; \mathrm{ad}P)$. The action of $\mathrm{Aut}(P)$ on $\mathcal{A}(P)$ is an affine action compatible with this linear action in the sense that:

$$\sigma^*(t\omega_0 + (1-t)\omega_1) = t\sigma^*\omega_0 + (1-t)\sigma^*\omega_1, \tag{5}$$

or equivalently,

$$\sigma^*(\omega + \pi^*\delta) = \sigma^*\omega + \pi^*(\rho(\sigma)^*(\delta)). \tag{6}$$

Let us consider the stabilizers of various points in $\mathcal{A}(P)$ under this action.

Lemma 4.3.3. *The stabilizer in the space of sections of* $\mathrm{Ad}P$ *of a connection* A *consists of all sections* $\sigma \colon B \to \mathrm{Ad}P$ *which are horizontal with respect to* A.

Proof. It suffices to work locally. We make a computation locally using a trivialization of $P|_U$ where U has local coordinates $(x_1, \ldots, x_n)$. The connection one-form restricted to the section $U \times \{1\}$ with respect to this trivialization is given by $\sum_i \alpha_i dx_i$. By choosing the local trivialization correctly we can suppose that $\alpha_1 = 0$. With respect to the local trivialization σ becomes a function from U to G. The formula in these local coordinates for $\sigma^*\omega$ is

$$\sigma^{-1}d\sigma + \sum_i \sigma^{-1}\alpha_i\sigma dx_i.$$

(See Problem 2.) The fact that $\sigma^*\omega = \omega$ then implies that

$$\sigma^{-1}\frac{\partial\sigma}{\partial x_1} = 0,$$

i.e., that

$$\frac{\partial\sigma}{\partial x_1} = 0.$$

This is exactly the condition that σ is horizontal with respect to the connection in the x_1-direction. Since the x_1-direction could have been chosen to be any direction at any point, it follows that σ is horizontal with respect to the connection.

The converse is proved similarly. □

Proposition 4.3.4. *If a section* σ *of* $\mathrm{Ad}P$ *stabilizes a connection* A*, then for each* $b \in B$ *we can write* $\sigma(b) = [p, h]$ *where* $h \in G$ *commutes with all elements in the image of the holonomy representation* $h_{A,p} \subset G$*. Furthermore, if* B *is connected, then the restriction map to a fiber induces an isomorphism between the stabilizer of a connection and the subgroup of* G *which centralizes the holonomy image at that point.*

Proof. First notice that if the first statement is true for one $p \in \pi^{-1}(b)$, then it is true for all $p' \in \pi^{-1}(b)$ since replace p by $p' = p \cdot g$ replaces h by $g^{-1}hg$ and replaces the image of $h_{A,p}$ by the image of $h_{A,pg}$ which is its conjugate by g.

Now let us suppose that σ stabilizes A. Consider a loop $\gamma(t)$ in B based by b. We write $\sigma(t)$ for the value of $\sigma(\gamma(t))$. We can write $\sigma(t) = [\tilde{\gamma}(t), g(t)]$ where $\tilde{\gamma}(t)$ is the horizontal lifting of $\gamma(t)$ beginning at p and $g(t) \in G$ is a smooth function of t. The fact that σ is A horizontal means that $g(t)$ is constant – in particular, $g(1) = g(0)$. But by definition of the holonomy map we have

$$\tilde{\gamma}(1) = \tilde{\gamma}(0) \cdot h_{A,p}(\gamma).$$

On the other hand, we know that $[\tilde{\gamma}(1), g(1)] = [\tilde{\gamma}(0), g(0)]$ so that

$$(h_{A,p}(\gamma))^{-1} \cdot g(1) \cdot h_{A,p}(\gamma) = g(0).$$

Together, these say that $g(0)$ centralizes $h_{A,p}(\gamma)$. Since γ was an arbitrary loop based at b, this completes the proof of the first assertion in the proposition.

Now suppose that B is connected and $g \in G$ centralizes the holonomy image $h_{A,p}(\Omega(B, b))$. There is a unique horizontal section of $\mathrm{Ad}P$ which agrees with $[p, g]$ at b. The uniqueness is clear; the only thing to prove is that if λ_1 and λ_2 are smooth paths in B from b to b', then parallel translating $[p, g]$ along these two paths gives the same result. That in turn is equivalent to showing that if λ is a loop in B based at b, then parallel translating $[p, g]$ around λ brings us back to $[p, g]$. This follows from the same argument given above. □

Corollary 4.3.5. *If $G = SU(2)$, then there is a subgroup $\{\pm 1\} \subset \mathrm{Aut}(P)$ consisting of the constant sections through $\pm 1 \in SU(2)$. This subgroup stabilizes every connection on P. If a connection A has a larger stabilizer, then A is induced from a connection on an S^1-bundle. In this case the stabilizer is isomorphic to S^1 unless the connection A is induced from a $\mathbf{Z}/2\mathbf{Z}$-connection in which case its stabilizer is isomorphic to $SU(2)$.*

Proof. Any subgroup of $SU(2)$ with centralizer bigger than $\{\pm 1\}$ is contained in a circle subgroup of $SU(2)$. In this case its centralizer is exactly the circle subgroup that contains it, unless the subgroup is contained in the center of $SU(2)$, in which case its centralizer is all of $SU(2)$. □

4.4. The space of gauge equivalence classes of connections

Now we wish to form the quotient space of the space of connections by the action of the group of gauge transformations. It is natural to think of the points of the quotient space as gauge equivalence classes of connections. In order to show that there is a reasonable quotient space, we need to show that the action has local slices. To establish such a result requires more analysis than we have invoked to date. The easiest context in which to do that analysis is in terms of Sobolev completions.

We introduce the L^p_k-norms on the space of C^∞-sections of a vector bundle $W \to B$ with a positive definite inner product. We also suppose that B is compact and has a riemannian metric, and let

$$\nabla\colon \Omega^0(B; \otimes^r T^*B \otimes W) \to \Omega^0(B; \otimes^{r+1} T^*B \otimes W)$$

be covariant differential with respect to a connection on W and the Levi-Civita connection on T^*B. Then

$$\|s\|_{L^p_k} = \int_B \left(\|s(x)\|^p + \|\nabla(s(x))\|^p + \cdots + \|\nabla^k(s(x))\|^p\right)^{\frac{1}{p}} \mathrm{dvol}.$$

Here the norms that we are integrating are the point-wise norms on $\otimes^r T^*B \otimes W$ induced by the given inner products on T^*B and W.

The completion of the C^∞-sections with respect to this norm is called *the Sobolev space of L^p_k-sections.* One can view the objects in these completions as distribution-valued sections, or simply as sequences of C^∞-sections which are Cauchy with respect to the given norm. We denote this completion as

$$\Omega^0_{L^p_k}(B; W).$$

There are several basic properties of these spaces that make them especially useful in many contexts. They are summarized in the following statements.

Theorem 4.4.1. *(The Sobolev Multiplication Theorem) Suppose that the dimension of B is n. There is a continuous (i.e. bounded) linear multiplication mapping*

$$\Omega^0_{L^p_k}(B;W)\otimes\Omega^0_{L^{p'}_{k'}}(B;W')\to\Omega^0_{L^q_r}(B;W\otimes W')$$

provided that $r\le\min(k,k')$ and

$$r-n/q<k-n/p+k'-n/p'.$$

Theorem 4.4.2. *(The Sobolev Embedding Theorem) Suppose that the dimension of B is n. There is a continuous embedding of $\Omega^0_{L^p_k}(B;W)$ into the C^r sections of W provided that $r<k-n/p$. Provided that $k-n/p>k'-n/p'$ and $k\ge k'$ there is a bounded inclusion*

$$\Omega^0_{L^p_k}(B;W)\subset\Omega^0_{L^{p'}_{k'}}(B;W).$$

This inclusion is compact if $k'<k$.

We can now define the space of L^p_k-connections on a smooth principal bundle P. These are connections A which, with respect to smooth local trivializations of $P\to B$ over an open set $U\subset B$, are given by a one-form α which is in $L^p_k(U)$ and which transforms by the usual formula when we change from one local trivialization to another. Clearly, this property is invariant as we change from one smooth local trivialization to another (because of the multiplication theorems). We denote this space of connections by $\mathcal{A}_{L^p_k}(P)$. The curvature of such a connection is given locally by the formula

$$F_A=d\alpha+\frac{1}{2}[\alpha,\alpha].$$

Again by the multiplication theorems we see that $F_A\in L^p_{k-1}$, provided that $k-n/p\ge 0$.

We can also define the group of L^p_{k+1}-changes of gauge. This is easiest to do when the gauge group $G\subset GL(n,\mathbf{R})\subset M_{n\times n}(\mathbf{R})$ is a linear group. In this case the fiber bundle AdP embeds in the vector bundle

$$\xi=P\times_G M_{n\times n}(\mathbf{R})$$

where the action of G on the matrices is by conjugation. Thus, we can consider L^p_{k+1}-sections of ξ which are also sections of AdP. This simply means every point in this space is represented by a sequence of C^∞-sections of AdP, which under this embedding forms a Cauchy sequence of sections of ξ with respect to the L^p_{k+1}-norm. Equivalently, we can consider L^2_{k+1}-sections of ξ whose image lies in Ad(P). (This is easier to think about in the case when $k+1-(n/p)>0$ in which case the sections are automatically continuous.)

Lemma 4.4.3. *With notation and assumptions as in the previous paragraph, suppose that $k-(n/p)\ge 0$. (Recall that the dimension of the base manifold B is n.) Then the space of L^p_{k+1}-gauge transformations forms a Banach Lie group $\mathcal{G}(P)$ with Lie algebra the space of L^p_{k+1}-sections of the adjoint bundle* ad$P$. *This group acts smoothly (in the sense of Banach manifolds) on the space $\mathcal{A}_{L^p_k}(P)$ of L^p_k-connections on P.*

Proof. The general theory of infinite dimensional manifolds (see [**8**], for example) implies that the space of L^p_{k+1}-sections of a C^∞ locally trivial fiber bundle $E \to B$ form a Banach manifold whose tangent space at a section σ is the Banach space of L^p_{k+1}-sections of $\sigma^* T_v E$ where $T_v E$ denotes the relative tangent bundle of E over B (that is to say the vertical tangent bundle or the bundle of tangents along the fibers of E over B). In our case with σ being the trivial section of $\mathrm{Ad}P$, the bundle of tangents along the fibers is identified with $\mathrm{ad}P$. Multiplication of matrices induces a smooth map $\xi \times_B \xi \to \xi$ which restricts to the map

$$\mathrm{Ad}P \times_B \mathrm{Ad}P \to \mathrm{Ad}P$$

which is multiplication along the fibers. The Sobolev multiplication theorem together with the fact that $k + 1 - (n/p > 0$ implies that the multiplication map induces a continuous linear map

$$L^p_{k+1}(\xi) \times L^p_{k+1}(\xi) \to L^p_{k+1}(\xi)$$

which induces by restriction a smooth map $\mathcal{G}(P) \times \mathcal{G}(P) \to \mathcal{G}(P)$ between Banach manifolds. This of course is the multiplication mapping for the group structure on $\mathcal{G}(P)$. Clearly, the inverse map is induced by a smooth, fiberwise diffeomorphism of $\mathrm{Ad}P$ and hence, the inverse is a smooth map of Banach manifolds as well. This shows that $\mathcal{G}(P)$ is a Banach Lie group. Clearly, the induced Lie algebra structure on the space of L^p_{k+1}-sections of $\mathrm{ad}P$ is given by fiberwise bracket.

Lastly, we consider the action of $\mathcal{G}(P)$ on the space $\mathcal{A}_{L^p_k}(P)$ of L^p_k-connections. This (right) action

$$\mathcal{A}_{L^p_k}(P) \times \mathcal{G}(P) \to \mathcal{A}_{L^p_k}(P)$$

is given by

$$(A, \sigma) \mapsto \sigma^* A.$$

The fact that this is a smooth action in the category of Banach manifolds follows easily from the fact that there is a continuous multiplication $L^p_{k+1} \otimes L^p_k \to L^p_k$. □

Remark 4.4.4. The Banach spaces L^2_k are actually Hilbert spaces, so that the above lemma holds in the category of Hilbert manifolds and Hilbert Lie groups for $p = 2$.

Now we assume that the base manifold B is a 4-dimensional manifold and that $P \to B$ is a principal $SU(2)$-bundle. We denote the group of L^2_3-gauge transformations by $\mathcal{G}_3(P)$ and the space of L^2_2-connections by $\mathcal{A}_2(P)$. Thus, our objects are Hilbert manifolds not just Banach manifolds. (We have chosen to work with these particular Sobolev spaces for simplicity only, we could work with any Sobolev spaces for which the group of gauge transformations is a continuous group with at least two derivatives, the spaces of connections has one fewer derivative, and the two-forms have two fewer derivatives.)

Proposition 4.4.5. *There are local slices for the action*

$$\mathcal{A}_2(P) \times \mathcal{G}_3(P) \to \mathcal{A}_2(P).$$

That is to say for each $A \in \mathcal{A}_2(P)$ there is a subspace

$$U \subset \mathcal{A}_2(P)$$

invariant under the stabilizer $\mathrm{Stab}(A)$ *of A such that the natural map*

$$U \times_{\mathrm{Stab}(A)} \mathcal{G}_3(P) \to \mathcal{A}_2(P)$$

is a diffeomorphism onto an open subset of $\mathcal{A}_2(P)$.

Before we prove this proposition we need a bootstrap lemma.

Lemma 4.4.6. *Let* $\{A_n\}$ *and* $\{B_n\}$ *be sequences of points in* $\mathcal{A}_2(P)$ *converging to connections* A *and* B. *Suppose that* $\{\sigma_n\}$ *is a sequence of elements of* $\mathcal{G}_3(P)$ *satisfying*

$$\sigma_n^* A_n = B_n.$$

Then after passing to a subsequence we can assume that $\{\sigma_n\}$ *converges to* $\sigma \in \mathcal{G}_3(P)$. *This limit satisfies* $\sigma^* A = B$.

Proof. We will define the L_r^p-norms by using the Levi-Civita connection on the tangent bundle of B and using ∇_A on the vector bundles associated to P. This allows us to define L_1^p-norms for any p, L_2^p-norms for any $p \leq 3$ and an L_3^2-norm. (See Problem 7).

We write $A_n = A + a_n$ and $B_n = A + b_n$. Then the sequences $\{a_n\}$ and $\{b_n\}$ converge to, say a and b in L_2^2. We consider σ_n as a section of the vector bundle $\xi = P \times_G M_{n \times n}$. Let us also use a local trivialization of this bundle. In the region where the trivialization holds, we have the equation

$$\sigma_n^{-1} d\sigma_n + \sigma_n^{-1}(A + a_n)\sigma_n = A + b_n$$

or equivalently,

$$d\sigma_n = \sigma_n A + \sigma_n b_n - A\sigma_n - a_n\sigma_n.$$

We write this as

$$d\sigma_n + [A, \sigma_n] = \sigma_n b_n - a_n \sigma_n.$$

The term on the right-hand-side is $\nabla_A(\sigma_n)$ when we view σ_n as a section of ξ. Thus, we get an equation in which the terms have meaning independent of the local trivialization. This then is a global equation:

$$\nabla_A(\sigma_n) = b_n\sigma_n - a_n\sigma_n. \tag{7}$$

Since a_n and b_n converge in L_2^2, by the Sobolev embedding theorems they converge and hence are bounded in L^5. Since the σ_n are L_3^2-sections, by the Sobolev embedding theorem they are continuous. Since the image of σ_n lies in a compact fiber bundle inside ξ, it follows that the L^∞-norm of the σ_n is universally bounded independent of n. It follows from this, Equation 7, and the continuity of the Sobolev multiplication $L^\infty \otimes L_2^2 \to L^5$ that the $\|\nabla_A(\sigma_n)\|_{L^5}$ are bounded independent of n. Together with L^∞-bound on the σ_n, this implies that the $\|\sigma_n\|_{L_1^5}$ are bounded independent of n. Thus, by the compact Sobolev embedding theorem, after passing to a subsequence we can assume that σ_n converge weakly in L_1^5 to an element σ, i.e., σ is in L_1^5 and σ_n converges to σ in L^5. It follows that $\nabla_A(\sigma_n)$ converges weakly in L^5 to $\nabla_A(\sigma)$. By the multiplication theorems $b_n\sigma_n - a_n\sigma_n$ converges weakly in L^5 to $b\sigma - \sigma a$. Thus, by the uniqueness of weak limits, we have the limiting equation in L^5

$$\nabla_A(\sigma) = b\sigma - \sigma a.$$

We are now in position to start the bootstrapping argument.

Claim 4.4.7. *For* $r = 0, 1, 2, 3$ *we have that* σ *is an* L_r^{5-r}*-section of* ξ *and* σ_n *converges in* L_r^{5-r} *to* σ.

Proof. The proof is by induction on r. We have just established the claim for $r=0$. Suppose that we know it for some $r<3$ and let us establish it for $r+1$. We use the Sobolev multiplication $L_r^{5-r}\otimes L_2^2\to L_r^{4-r}$. From Equation 7 we see that since the σ_n converge to σ in L_r^{5-r} and the a_n and b_n converge in L_2^2 to a and b, it follows that $\nabla_A(\sigma_n)=b_n\sigma_n-\sigma_n a_n$ converges in L_r^{4-r} to $\nabla_A(\sigma)=b\sigma-\sigma b$. Together with the fact that σ has already been show to be the limit of the σ_n in L_r^{5-r}, this implies that σ is an L_{r+1}^{4-r} limit of the σ_n. This completes the induction and hence the proof of the claim. □

At this point we have shown that after passing to a subsequence the σ_n converge in L_3^2 to σ. In particular, the limit σ is in $\mathcal{G}_3(P)$. It is clear from the fact that $\nabla_A(\sigma)=b\sigma-\sigma a$ that $\sigma^*A=B$, establishing Lemma 4.4.6. □

Now we are in a position to prove the proposition.

Proof. We introduce a riemannian metric on B. We also assume that B is oriented. This induces a $*$-operator on forms and hence allows us to define the L^2-adjoints of our operators. First, let us consider the L^2-adjoint of ∇_A,

$$\nabla_A^*\colon \Omega^1_{L_2^2}(B;\operatorname{ad}P)\to\Omega^0_{L_1^2}(B;\operatorname{ad}P).$$

(This adjoint is in fact equal $-*(\nabla_A)*$.) One of the basic properties of these Sobolev spaces and the elliptic operators like ∇_A^* is that $\operatorname{Ker}(\nabla_A^*)$ is a closed subspace of $\Omega^1_{L_2^2}(B;\operatorname{ad}P)$ and that

$$\Omega^1_{L_2^2}(B;\operatorname{ad}P)=\nabla_A(\Omega^0_{L_3^2}(B;\operatorname{ad}P))\oplus\operatorname{Ker}(\nabla_A^*).$$

Of course, the linear subspace $\operatorname{Ker}(\nabla_A^*)$ is invariant under the action of $\operatorname{Stab}(A)$.

We have a smooth map of Hilbert manifolds

$$\operatorname{Ker}(\nabla_A^*)\times_{\operatorname{Stab}(A)}\mathcal{G}_3(P)\to\mathcal{A}_2(P)$$

given by

$$(u,\sigma)\mapsto\sigma^*(A+u).$$

A direct computation shows that its differential at $[1,0]$ is a linear isomorphism. Thus, by the inverse function theorem for Banach manifolds, this map induces a diffeomorphism from a neighborhood of $[0,1]$ to a neighborhood of A in $\mathcal{A}_2(P)$.

The slice $U\subset\mathcal{A}_2(P)$ is of the form

$$\{A\}+B$$

where B is the ball of radius $\epsilon>0$ in $\operatorname{Ker}(\nabla_A^*)$ for some ϵ sufficiently small. As we have just seen, if we choose $\epsilon>0$ sufficiently small, then the map

$$U\times_{\operatorname{Stab}(A)}\mathcal{G}_3(P)\to\mathcal{A}_2(P)$$

will be a local diffeomorphism onto an open neighborhood of the $\mathcal{G}_3(P)$-orbit through A. We must show that, possibly after replacing B by a smaller ball, this map is globally one-to-one. If there is no such smaller ball, then there are elements $\{u_n,v_n\}_{n=1}^\infty$ in U and elements $\sigma_n\in\mathcal{G}_3(P)$ such that

- $\sigma_n^*(A+u_n)=A+v_n$, and
- $\lim_{n\mapsto\infty}u_n=\lim_{n\mapsto\infty}v_n=0$.
- $[u_n]\neq[v_n]\in U/\operatorname{Stab}(A)$.

According to the previous lemma, after passing to a subsequence we can extract a limiting gauge transformation σ of the σ_n. It stabilizes A. Thus, the sequences $[b_n, 1]$ and $[a_n, \sigma_n]$ in $U \times_{\mathrm{Stab}(A)} \mathcal{G}_3(P)$ both converge to $[0, 1]$. By the local diffeomorphism property this implies that for all $n >> 0$ we have $[b_n, 1] = [a_n, \sigma_n]$, that is to say we have $\sigma_n^* a_n = b_n$. This contradicts the third item in the statement of the proposition. □

4.5. The local structure of the quotient space

Using the slice theorem and the list of possible stabilizers of connections we can describe the local structure of the space $\mathcal{B}_2(P) = \mathcal{A}_2(P)/\mathcal{G}_3(P)$ of gauge equivalence classes of connections on a principal $SU(2)$-bundle.

First of all it follows immediately from Lemma 4.4.6 that $\mathcal{B}_2(P)$ is a Hausdorff space. For suppose that $[A]$ and $[B]$ are points of $\mathcal{B}_2(P)$ with the property that every neighborhood of $[A]$ meets every neighborhood of $[B]$. Then we can find sequences of connections A_n converging to A and B_n converging to B such that $[A_n] = [B_n]$. The last condition means that there are elements $\sigma_n \in \mathcal{G}_3(P)$ with $\sigma^* A_n = B_n$. By the lemma we can extract a convergent subsequence of the σ_n converging to σ with $\sigma^* A = B$. Hence, $[A] = [B]$.

Provided that $c_2(P) \neq 0$ then the only possible stabilizers for elements of $\mathcal{A}_2(P)$ are $\{\pm 1\}$ or S^1. We call the former connections *irreducible* and the latter ones *reducible*; and we denote by $\mathcal{A}_2^*(P)$ the space of irreducible connections. It is an open subset of $\mathcal{A}_2(P)$ and hence is a Hilbert manifold. For the induced action of $\mathcal{G}_3(P)$ on $\mathcal{A}_2^*(P)$, the stabilizer of every point is $\{\pm 1\}$. Thus, we have a free action of $\mathcal{G}_3(P)/\{\pm 1\}$ on $\mathcal{A}_2^*(P)$. According to Proposition 4.4.5 this action has slices. The slice through a connection $A \in \mathcal{A}_2^*(P)$ is modeled on $\mathrm{Ker}\nabla_A^*$. It follows that the quotient space $\mathcal{B}_2^*(P) = \mathcal{A}_2^*(P)/\mathcal{G}_3(P)$ is a Hilbert manifold. Its tangent space at $[A]$ is identified with $\Omega^1_{L^2_2}(B; \mathrm{ad}P)/\mathrm{Im}\nabla_A = \mathrm{Ker}\nabla_A^*$.

If A is a reducible connection with stabilizer S^1, then a neighborhood of $[A] \in \mathcal{B}_2(P)$ is homeomorphic to U/S^1. The element $-1 \in S^1$ fixes every connection, but the quotient circle obtained by dividing out by this element acts semi-freely on U. The fixed point set is the subset of reducible connections in U. Since the action on U is linear, we can decompose the linear space $\mathrm{Ker}\nabla_A^*$ as $\mathbf{H_R} \oplus \mathbf{H_C}$, the sum of a real and a complex Hilbert space. The action of S^1 on this space is the square of complex multiplication on $\mathbf{H_C}$. Thus, the quotient is locally isomorphic to

$$\mathbf{H_R} \times \left((\mathbf{H_C}) / S^1 \right).$$

4.6. Problems

1. Show that if $E = P \times_G F$ is a locally trivial fiber bundle associated to a principal G-bundle P, then the natural action of $\mathrm{Aut}(P)$ on P induces an action on E.
2. Show that if $P \to B$ has a trivialization and if the restriction of a connection one-form A from P to the section $B \times \{1\}$ is $\alpha \in \Omega^1(B; \mathfrak{g})$, then the restriction of the connection $\sigma^* A$ to this section is given by

$$\sigma^{-1} d\sigma + \sigma^{-1} \alpha \sigma.$$

3. (a) Compute the possible stabilizers for connections on principal $SO(3)$-bundles. (b) Do the same for principal $SU(3)$-bundles.
4. Show that if $k > n/2$ then any L^2_k-section of a vector bundle over an n-manifold is continuous. In particular, the group of L^2_3-gauge transformations on a principal G-bundle over a 4-manifold is a group of continuous bundle automorphisms.
5. For a 4-dimensional manifold base show that the group of L^2_{k+1}-changes of gauge acts smoothly on the affine space of L^2_k-connections, provided that $k \geq 2$.
6. Show that $\mathcal{A}^*(P)$ of irreducible $SU(2)$-connections is contractable. [Hint: It is the complement of a subset of infinite codimension in an affine Banach manifold.]
7. Let B have a riemannian metric and let ∇_A^* be the L^2-adjoint of ∇_A. Show that the tangent space to $\mathcal{B}^*(P)$ at $[A]$ can be identified with $\mathrm{Ker}\nabla_A^*$. This determines a riemannian metric on $\mathcal{B}^*(P)$.
8. Show that $\mathcal{B}^*(P)$ is a classifying space for the group $\mathcal{G}(P)/\{\pm 1\}$.
9. Show that if A is an L^2_2-connection, then on any associated vector bundle W we can define an L^2_3-norm of sections by using ∇_A. Similarly, show that A can be used to define an L^3_2-norm, and an L^p_1-norm for any $p > 0$.
10. Compute the local structure of $\mathcal{B}(P)$ near a flat connection with holonomy at every point of B contained in the center $\{\pm 1\}$ of $SU(2)$.
11. Show that the function
$$d\colon \mathcal{A}(P) \times \mathcal{A}(P) \to \mathbf{R}$$
given by
$$d(\omega_1, \omega_2) = \inf_{\sigma \in \mathcal{G}} \|\omega_1 - \sigma^*(\omega_2)\|_{L^2}$$
is $\mathcal{G}$-invariant, hence symmetric in the two factors.
12. Show that the map induced by d on $\mathcal{A}(P)/\mathcal{G}(P) \times \mathcal{A}(P)/\mathcal{G}(P)$ is actually a metric, called the L^2-metric.
13. Viewing curvature as a function
$$F\colon \mathcal{A}(P) \to \Omega^2(B; \mathrm{ad}P),$$
show that its differential at a connection A, viewed as a map
$$DF\colon \Omega^1(B; \mathrm{adP}) = \mathrm{T_A}(\mathcal{A}(\mathrm{P})) \to \Omega^2_{\mathrm{L}^2_1}(\mathrm{B}; \mathrm{adP}),$$
is equal to
$$DF = \nabla_A,$$
the induced covariant derivative.
14. Show that the curvature function F induces a well-defined section of $\mathcal{A}(P) \times_{\mathcal{G}(P)} \Omega^2_{L^2_1}(B; \mathrm{adP})$, which is a Banach vector bundle over $\mathcal{B}(P)$.
15. Fix a connection A. This induces a map
$$\mathcal{G}(P) \to \mathcal{A}(P)$$
by
$$\sigma \mapsto \sigma^*(A).$$
Show that the derivative of this map is given by
$$\nabla_A\colon \Omega^0(B; \mathrm{adP}) \to \Omega^1(B; \mathrm{adP}).$$

16. Let $A \in \mathcal{A}(P)$ be a connection, $a \in \Omega^1(B; \mathrm{adP})$, write down the Euler-Lagrange equation for the functional $\mathcal{A}(P) \to \mathbf{R}$ given by
$$C \mapsto \|A - C\|_{L^2}$$
restricted to the $\mathcal{G}(P)$-orbit through $A + a$. More precisely, show that the condition on a that for any $\xi \in T_e\mathcal{G}(P) = \Omega^0(B; \mathrm{adP})$,
$$\frac{d}{dt}\|exp(t\xi)^*(A + a) - A\|_{L^2} = 0$$
is that
$$\nabla_A^*(a) = 0.$$
If $C = A + a$ satisfies this equation, C is said to be in *Coulomb gauge* respect to A. Relate the Coulomb gauge condition to the local slices discussed in the text.
17. Show that the Coulomb gauge relation is symmetric; i.e. A is in Coulomb gauge with respect to C iff C is in Coulomb gauge with respect to A.
18. Show that any $SU(2)$-bundle over a three-manifold is trivial.
19. Show that the normalized second Chern form is given by
$$\frac{1}{8\pi^2} Tr(F_A \wedge F_A).$$
By $Tr(F_A \wedge F_A)$ we mean the following. Evaluating F_A on a pair of vectors produces an element in $\mathfrak{g}$. We evaluate both curvature terms to produce two elements in $\mathfrak{g}$. Unlike the situation before, we do not take the Lie bracket of these elements. Rather we take the trace of their product.
20. We know from Chern-Weil theory that if B is a closed four-manifold, and A is a connection on an $SU(2)$ bundle $P \to B$, then
$$\frac{1}{8\pi^2}\int_B Tr(F_A \wedge F_A) = \int_B c_2(P).$$
Moreover, it is a classical fact that any oriented three-manifold bounds an oriented four-manifold. Take an oriented three-manifold N, a principal $SU(2)$-bundle over it Q, and a connection over that bundle A. Show that there always is some oriented four-manifold M with $SU(2)$ bundle P and connection $\widetilde{A}$ with the property that $\partial M = N$, with $(P|_N, \widetilde{A}|_N) = (Q, A)$. Show that the real number
$$\frac{1}{8\pi^2}\int_B Tr(F_{\widetilde{A}} \wedge F_{\widetilde{A}}),$$
depends modulo $\mathbf{Z}$ only on the bounding data (N, Q, A), inducing a map
$$CS\colon \mathcal{A}(N \times SU(2)) \to \mathbf{R}/\mathbf{Z}.$$
21. Show that the above map descends to give a map
$$CS\colon \mathcal{B}(N \times SU(2)) \to \mathbf{R}/\mathbf{Z},$$
the so-called *Chern-Simons functional.*
22. Show that
$$CS(A) \equiv 0 \pmod{\mathbf{Z}}$$
when A is the product connection.

23. We would like to understand this functional explicitly. If a is the "connection matrix" for A, i.e. $A = \nabla + a$, where ∇ is the product connection, show that
$$CS(A) = \frac{1}{8\pi^2}\int_B Tr(a \wedge da + \frac{2}{3} a \wedge a \wedge a).$$
More generally, if A is a connection, show that
$$CS(A+a) = CS(A) + \frac{1}{8\pi^2}\int_B Tr(2F_A \wedge a + a \wedge \nabla_A a + \frac{2}{3} a \wedge a \wedge a).$$
Hint: Consider the cobordism $N \times I$, with the connection $\frac{\partial}{\partial t} + A + at$, where t is the coordinate in the I direction. Now show that $CS(A+a) - CS(A)$ is computed by integrating the curvature of this connection over the tube, and then perform the integral.
24. Use the formula from Problem 23 to show that the critical points of the Chern-Simons functional are the flat connections.
25. Show that any element of $su(2)$, the Lie algebra of $SU(2)$, can be written as
$$\begin{pmatrix} ia & b+ic \\ -b+ic & -ia \end{pmatrix}.$$
This gives an identification of $su(2)$ with $\mathbf{R}^3$. Thus, the standard metric on $\mathbf{R}^3$ induces a metric on $su(2)$ (the square of the distance function being $a^2+b^2+c^2$). Compare this metric with the function
$$\mathfrak{g} \times \mathfrak{g} \to \mathbf{R}$$
given by $(A, B) \mapsto Tr(AB)$. Show that with respect to this induced metric on $su(2)$ and the L^2 metric on $\mathcal{B}(N \times SU(2))$, the gradient of CS at the connection A is given by
$$- * F_A \in \Omega^1(N, \mathrm{adP}).$$

LECTURE 5
The ASD Equations and the Moduli Space

In this section we continue the notation and conventions from the previous lecture: B is a smooth, oriented, compact 4-manifold, $P \to B$ is a principal $SU(2)$-bundle, and B has a riemannian metric.

5.1. The ASD equations and the moduli space

The riemannian metric on B induces the Hodge $*$-operator

$$*\colon \Omega^i(B;V) \to \Omega^{4-i}(B;V)$$

for any vector bundle V over B. This operator satisfies $*^2 = (-1)^{i\cdot(4-i)}$. In particular, the $*$ operator restricted to 2-forms is an involution. It determines a decomposition

$$\Omega^2(B;V) = \Omega^2_+(B;V) \oplus \Omega^2_-(B;V)$$

of the two-forms into the $+1$ and -1 eigenspaces for $*$. These subspaces are the subspaces of self-dual and anti-self-dual two-forms. An ASD (SD) connection is one A whose curvature $F_A \in \Omega^2(B; \mathrm{ad}P)$ is anti-self-dual (self-dual). If we reverse the orientation of B then this changes SD connections into ASD connections and vice versa. Thus, the two theories are completely equivalent. From this one might be tempted to conclude that there is no reason to introduce and extra sign, and therefore one should simply work with SD connections. But there is one important class of 4-manifolds which have a natural orientation – complex surfaces. Working with the natural orientation of these manifolds, it turns out that ASD connections are closely related to certain algebro-geometric objects (stable holomorphic vector bundles) associated to the surfaces whereas SD connections are related to anti-holomorphic objects. For this reason we cast the whole theory from the beginning in terms of ASD connections.

It is clear that the subset $\widetilde{\mathcal{M}}(P) \subset \mathcal{A}_2(P)$ of all ASD connections is invariant under the action of $\mathcal{G}_3(P)$. We denote the quotient

$$\mathcal{M}(P) = \widetilde{\mathcal{M}}(P)/\mathcal{G}_3(P).$$

We also have

$$\mathcal{M}^*(P) = \mathcal{M}(P) \cap \mathcal{B}_2^*(P).$$

We call these respectively the *moduli space of ASD connections* and *moduli space of irreducible ASD connections*. Later, we will be considering moduli spaces with respect to different metrics. When the metric we are using to define the moduli space is not clear for the context we will also denote these spaces by $\mathcal{M}(P,g)$ and $\mathcal{M}^*(P,g)$, thus indicating the metric being used.

5.2. The local structure of the moduli space

The purpose of this paragraph is to study the basic properties of this moduli space. We would like to have some analogue of the implicit function theorem which would describe this space. We have the self-dual part of the curvature which defines a smooth mapping

$$P_+F\colon \mathcal{A}_2(P) \to \Omega^2_+(B;\mathrm{ad}P)_{L^2_1}.$$

The preimage of zero under this mapping is the space $\widetilde{\mathcal{M}}(P)$ of ASD connections on P. Thus, the setup must involve some infinite dimensional version of the Implicit Function Theorem. The one that we will use is the following:

Lemma 5.2.1. *Suppose that $F\colon A \to B$ is a smooth map between Hilbert manifolds whose differential at every point $a \in A$, DF_a, is a Fredholm map $TA_a \to TB_{f(a)}$. Then near each point of $a \in F^{-1}(b)$ there is a finite dimensional smooth model describing $F^{-1}(b)$ locally. That is to say: There are neighborhoods $U \subset A$ of a and $V \subset B$ of b and decompositions $U = U_1 \times U_2$ and $V = V_1 \times V_2$ with the following properties:*

- $b = (b_1, b_2)$,
- *U_2 and V_2 are finite dimensional manifolds,*
- *$(u_1,u_2) = (F_1(u_1), F_2(u_1,u_2))$ where $F_1\colon U_1 \to V_1$ is a diffeomorphism.*

It follows immediately that a neighborhood of $a \in F^{-1}(b)$ is isomorphic $\Phi^{-1}(b_2)$ where $\Phi\colon U_2 \to V_2$ is defined by $\Phi(u_2) = F_2(F_1^{-1}(b_1), u_2)$.

Proof. See any standard reference on infinite dimensional manifolds. □

This then shows that the preimage of zero under a smooth map with Fredholm differential at each point is locally isomorphic to a subset defined by the zeros of a smooth map between finite dimensional smooth manifolds. The nicest case is the following one.

Corollary 5.2.2. *If DF_a is surjective, then a neighborhood of a in $F^{-1}(b)$ is a smooth manifold of dimension equal to the dimension of* $\mathrm{Ker}DF_a$.

Corollary 5.2.3. *If DF_a is surjective for every $a \in F^{-1}(b)$, then $F^{-1}(b)$ is a smooth finite dimensional manifold of dimension equal to the dimension of* $\mathrm{Ker}DF_a$ *for any $a \in F^{-1}(b)$.*

We wish to apply this theory to the moduli space. Unfortunately, this theory does not apply to the curvature map

$$P_+F\colon \mathcal{A}_2(P) \to \Omega^2_+(B;\mathrm{ad}P)_{L^2_1}$$

as the next lemma shows.

Lemma 5.2.4. *The differential of the self-dual curvature P_+F at an ASD connection A is*

$$P_+\nabla_A\colon \Omega^1_{L^2_2}(B;\mathrm{ad}P) \to \Omega^2_+(B;\mathrm{ad}P)_{L^2_1}.$$

The proof of this lemma is straightforward and is left as an exercise (Problem 36).

The reason is that the Fredholm theory does not apply to this map is that $P_+\nabla_A$ is not a Fredholm operator. Indeed, since A is ASD, we know that $P_+\nabla_A \circ \nabla_A = 0$, and hence the kernel of $P_+\nabla_A$ contains the infinite dimensional subspace $\mathrm{Im}\nabla_A$. One sees this same infinite dimensionality reflected at the non-linear level – since $F^{-1}(0) = \widetilde{\mathcal{M}}(P)$ has a free action of an infinite dimensional group, it can not have local finite dimensional models. The Fredholm theory only applies when we work with the quotient of the group of gauge transformations. One way to do this is to restrict to the slices.

Lemma 5.2.5. *Let A be an ASD connection. The restriction of the curvature map to the slice* $\mathrm{Ker}\nabla_A^*$

$$P_+\nabla_A\colon \mathrm{Ker}\nabla_A^* \to \Omega^2_+(B;\mathrm{ad}P)_{L^2_1}$$

is a Fredholm linear operator.

Proof. This is a standard result in the theory of elliptic operators. Consider the following sequence of operators

$$\mathcal{E}(A):\qquad 0 \to \Omega^0(B;\mathrm{ad}P) \xrightarrow{\nabla_A} \Omega^1(B;\mathrm{ad}P) \xrightarrow{P_+\nabla_A} \Omega^2_+(B;\mathrm{ad}P) \to 0.$$

If A is ASD, then the sequence is a complex (in general, $P_+\nabla_A\circ\nabla_A$ is multiplication by P_+F_A). The associated symbol complex is a complex of vector bundles and linear maps over it cotangent bundle T^*B of B.

$$0 \to p^*(\wedge^0 T^*B \otimes \mathrm{ad}P) \xrightarrow{\eta_1} p^*(\wedge^1 T^*B \otimes \mathrm{ad}P) \xrightarrow{\eta_2} p^*(\wedge^2_+ T^*B \otimes \mathrm{ad}P) \to 0.$$

Here $p\colon T^*B \to B$ is the projection mapping. The symbol maps on the vector spaces in these bundles over the cotangent vector x are

$$\eta_1|_x = (\wedge x)\otimes \mathrm{Id}_{\mathrm{ad}P_{p(x)}}$$

and

$$\eta_2|_x = P_+((\wedge x)\otimes \mathrm{Id}_{\mathrm{ad}P_{p(x)}}).$$

The exactness of this sequence for $x \neq 0$ follows immediately from the analogous statement where the bundle $\mathrm{ad}P$ is replaced by the trivial bundle $\mathbf{R}$. Exactness for this sequence is well-known and is a straightforward computation in linear algebra (see Problem 13).

Standard elliptic theory now tells us that in the original complex the images of the operators are closed subspaces and the cohomology groups are finite dimensional. In particular, the image of $P_+\nabla_A$ is closed and the cokernel of this map is finite dimensional. It also gives us a decomposition

$$\Omega^1_{L^2_2}(B;\mathrm{ad}P) = \nabla_A(\Omega^0_{L^2_3}(B;\mathrm{ad}P)) \oplus \mathrm{Ker}\nabla_A^*.$$

Thus, $\mathrm{Ker}(P_+\nabla_A|_{\mathrm{Ker}\nabla_A^*})$ is identified with the first cohomology group of the complex, and hence is finite dimensional. □

The complex $\mathcal{E}(A)$ is called the *elliptic complex determined by the ASD connection A*. Notice that the three spaces in this complex are respectively the Lie algebra of the group of gauge transformations, the tangent space to the space of connections at A, and the space of self-dual two-forms. The first linear operator is the differential of the action of $\mathcal{G}_3(P)$ on $\mathcal{A}_2(P)$ at the identity of $\mathcal{G}_3(P)$ and A, thus, the image of the first operator is the tangent space to the $\mathcal{G}_3(P)$-orbit through

A. The second operator is the differential of the curvature mapping at A. Thus, the kernel of the second operator at an ASD connection A is the formal tangent space to the space $\widetilde{\mathcal{M}}(P)$ at the point A. Hence, the first cohomology $H^1(\mathcal{E}(A))$ is the formal tangent space to $\mathcal{M}(P)$ at $[A]$. In fact, according to the Implicit Function Theorem there is a mapping

$$\Psi\colon H^1(\mathcal{E}(A)) \to H^2(\mathcal{E}(A))$$

whose differential at 0 is trivial, so that $\Psi^{-1}(0)$ is identified with a neighborhood of $[A]$ in $\mathcal{M}^*(P)$ provided that A is irreducible. (If A is reducible, then $\Psi^{-1}(0)/\mathrm{Stab}(A)$ is homeomorphic to a neighborhood of $[A]$ in $\mathcal{M}(P)$.) This map is called the *Kuranishi map* or the *Kuranishi obstruction map*. The nicest case is when in addition the second cohomology $H^2(\mathcal{E}(A))$ is zero, in which case the moduli space is a smooth manifold near $[A]$ of dimension equal to the dimension of H^1 of the complex, which in turn is equal to minus the Euler characteristic of the basic elliptic complex.

This Euler characteristic is independent of the connection A and can be calculated via the Atiyah-Singer Index Theorem in terms of the characteristic classes of $\mathrm{ad}P$ and homotopy invariants of B. It turns out that the formula is:

$$\chi(\mathcal{E}(A)) = 3(b_0(B) - b_1(B) + b_2^+(B)) - 8c_2(P)$$

where the $b_i(B)$ are the Betti numbers of B and $b_2^+(B)$ is the dimension of a maximal positive definite subspace of $H^2(B;\mathbf{R})$ under the intersection pairing. This formula is a consequence of the Atiyah-Singer Index theorem.

Corollary 5.2.6. *If A is an irreducible ASD connection and if*

$$P_+\nabla_A\colon \Omega^1_{L^2_2}(B;\mathrm{ad}P) \to \Omega^2_+(B;\mathrm{ad}P)_{L^2_1}$$

is surjective, then $[A]$ is a smooth point $\mathcal{M}^*(P)$. Near this point $\mathcal{M}^*(P)$ is a smooth submanifold of $\mathcal{B}^*_2(P)$ of dimension

$$8c_2(P) - 3(b_0(B) - b_1(B) + b_2^+(B)).$$

Near $[A]$ the moduli space is cut out transversely by the self-dual curvature in the following sense. The self-dual curvature gives a smooth Fredholm section of the vector bundle

$$\mathcal{A}^*_2(P) \times_{\mathcal{G}_3(P)} \Omega^2_+(B;\mathrm{ad}P)_{L^2_1}.$$

The subspace of zeros of this section is exactly $\mathcal{M}^*(P)$. The section is transverse to the zero section near $[A]$.

We also have a result near reducible connections if $H^2(\mathcal{E}(A))$ is trivial.

Corollary 5.2.7. *Suppose that A is a reducible ASD connection whose stabilizer is S^1. Suppose also that $P_+\nabla_A$ is surjective. Then there is a neighborhood $\mathcal{O}$ of $[A] \in \mathcal{M}(P)$ homeomorphic to $\mathbf{R}^k \oplus \mathbf{C}^n/S^1$ where the S^1 acts via complex multiplication. The subspace $\mathbf{R}^k \times \{0\}$ corresponds under this homeomorphism to the subspace of classes of reducible ASD connections in $\mathcal{O}$.*

Proof. Let U be a slice for the $\mathcal{G}_3(P)$-action at A. Our hypothesis is that

$$P_+\nabla_A\colon \mathrm{Ker}\nabla^*_A \to \Omega^2_+(B;\mathrm{ad}P)_{L^2_1}$$

is surjective. We know that it is Fredholm. Thus, the preimage $(P_+F_A)^{-1}(0) \cap U$ is a smooth manifold. This manifold is invariant under the action of S^1 which is semi-free (after dividing out by $\{\pm 1\}$ which acts trivially). The result is now clear. □

5.3. The generic metrics theorem

So far we have always fixed the riemannian metric on B and worked with the ASD equations relative to that metric. Of course, the $*$-operator, and hence the moduli space $\mathcal{M}^*(P)$, depend on the choice of metric. Any given metric may not yield a good moduli space. We may have to vary the metric in order to find a moduli space with the best possible properties. It is the purpose of this section to outline how we show that such good metrics exist. We proceed by studying all metrics at once and trying to understand which are 'good' for our purposes.

Fix $r \geq 3$ and consider the space $\mathcal{C}$ of C^r-riemannian metrics on B. This is naturally a Banach manifold with the C^r-norm. Here is the generic metrics theorem.

Theorem 5.3.1. *For any closed, oriented 4-manifold B and any principal $SU(2)$-bundle $P \to B$ whose Chern class is non-zero, there is a dense G_δ-subset of the space of metrics $\mathcal{C}$ with the following property. If g is in the G_δ-subset then for every g-ASD connection A the operator*

$$P_+\nabla_A \colon \Omega^1_{L^2_2}(B; \mathrm{ad}P) \to \Omega^2_+(B; \mathrm{ad}P)_{L^2_1}$$

is surjective. It follows that the associated moduli space $\mathcal{M}^*(P, g)$ is a smooth submanifold of $\mathcal{B}^*(P)$.

We will not give a complete proof of this result here, but merely sketch it. A complete proof can be found in [4].

First we form the parametrized space of connections $\mathcal{A}_2(P) \times \mathcal{C}$. Over this space there is a vector bundle which we denote $\widetilde{\Omega}^+$, whose fiber over (A, g) is the vector space of g-self-dual two-forms on B with values in ad$P$. This bundle sits inside the product bundle

$$(\mathcal{A}_2(P) \times \mathcal{C}) \times \Omega^2(B; \mathrm{ad})P_{L^2_1}.$$

Associating to (A, g) the g-self-dual part of the curvature of A defines a smooth section P_+F_A of $\widetilde{\Omega}^+$. The most difficult part of the argument technically is showing that this section has surjective differential at every zero, i.e., that this section is transverse to the zero section. This is the only place that one uses the fact that the bundle P is non-trivial.

The group $\mathcal{G}_3(P)$ naturally acts linearly on the total space of $\widetilde{\Omega}^+$ and the subgroup $\{\pm 1\}$ acts trivially. Thus, we can form the quotient by this action of the restriction of $\widetilde{\Omega}^+$ over the subspace $\mathcal{A}^*_2(P) \times \mathcal{C}$. The quotient is a vector bundle Ω^+ over $\mathcal{B}^*_2(P) \times \mathcal{C}$. By the surjectivity described in the previous paragraph, we see that the self-dual part of the curvature P_+F_A defines a smooth section of Ω^+ which is transverse to the zero section. Its zero locus is the parametrized moduli space of irreducible ASD connections $\mathcal{M}^*(P, \mathcal{C})$ consisting of all pairs $([A], g)$ where A is an irreducible g-ASD connection. It follows that this parametrized moduli space is an infinite dimensional smooth submanifold of $\mathcal{B}^*_2(P) \times \mathcal{C}$. Next, one shows that the projection mapping $\mathcal{M}^*(P, \mathcal{C}) \to \mathcal{C}$ is a smooth map with Fredholm differential of index equal to

$$d(P) = 8c_2(P) - 3(b_0(B) - b_1(B) + b_2^+(B)). \tag{8}$$

The standard Banach space version of Sard's theorem can now be invoked to show that for a generic point $g \in \mathcal{C}$ the intersection

$$\mathcal{M}^*(P,g) = (\mathcal{B}_2^*(P) \times \{g\}) \cap \mathcal{M}^*(P,\mathcal{C})$$

is a smooth submanifold of dimension $d(P)$ in the sense that the section P_+F_A restricted to this slice has surjective differential at its zeros. Consequently, for any such metric g the moduli space $\mathcal{M}^*(P,g)$ is a smooth manifold cut out transversely by P_+F_A. To extend this result to reducible connections, one considers based connections.

5.4. Reducible connections

There is another issue besides the smoothness of $\mathcal{M}^*(P)$ that we must deal with. That is the question of whether there are reducible ASD connections. Let us give a criterion when these exist.

Lemma 5.4.1. *Let us fix a riemannian metric g on B. The bundle P admits a reducible ASD connection whose associated principal S^1-bundle has first Chern class equal to $x \in H^2(B;\mathbf{Z})$ if and only if the harmonic form representative for x with respect to g is an ASD two-form.*

Proof. First suppose that the harmonic form representative h for x is ASD. Let $Q \to B$ be the principal S^1-bundle whose first Chern class is x. There is a connection on Q whose curvature form is h (see Problem 35). Obviously, this connection is ASD. Now we form the induced connection on $P = Q \times_{S^1} SU(2)$. It is easy to see that the curvature of this induced $SU(2)$-connection is also ASD (see Problem 34). This proves one direction of the lemma.

Suppose on the other hand that P has a reducible ASD connection A. This connection determines a reduction of P to some principal S^1-bundle Q. The connection A determines an ASD connection on Q (see Problem 34) whose curvature is an ASD closed form representing $c_1(Q)$. Since the curvature is closed and ASD, it is harmonic (see Problem 37). □

The following result shows that in most contexts one can assume that there are no reducible ASD connections.

Theorem 5.4.2. *The subset in the space of metrics $g \in \mathcal{C}$ for which there exist reducible g-ASD connections is a countable union of smooth, locally closed submanifolds of codimension $b_2^+(B)$.*

Proof. We sketch a proof of this result. A complete proof can be found in [9] or [3]. First let us fix a class $x \in H^2(B;\mathbf{Z})$ with $x^2 = -c_2(P)$. Let Q be the principal S^1-bundle with first Chern class x. There is an isomorphism of principal $SU(2)$-bundles $Q \times_{S^1} SU(2) \to P$, i.e., there is a reduction of P with underlying S^1-bundle Q. We wish to understand the subspace of $R_x \subset \mathcal{C}$ consisting of all metrics for which there is a reducible ASD connection which gives this topological reduction.

As the previous lemma shows, this is the subspace of metrics for which the harmonic representative for x is ASD. Let N be the grassmannian of linear subspaces of $H^2(B;\mathbf{R})$ of dimension $b_2^-(B)$. Consider the smooth map $n\colon \mathcal{C} \to N$ which associates to g the subspace of $H^2(B;\mathbf{R})$ represented by ASD g-harmonic forms. Letting $C_x \in N$ be the subset of planes that contain x, it is an easy exercise to show that C_x is a smooth submanifold of N of codimension $b_2^+(B)$. Of course, $R_x = n^{-1}(C_x)$. It is true (see [9]) that this map has surjective differential although we will not supply the proof. Given this, it is immediate that R_x is a smooth closed submanifold of $\mathcal{C}$ of codimension $b_2^+(B)$.

Since there are only countably many $x \in H^2(B;\mathbf{Z})$ with $x^2 = -c_2(P)$, the result is clear. □

Putting together the main results of the last two paragraphs gives us the following result.

Theorem 5.4.3. *Let B be a closed, oriented four-manifold with $b_2^+(B) > 0$. There is a dense G_δ in the space $\mathcal{C}$ of riemannian metrics on B so that if g is in this subset then $\mathcal{M}^*(P,g) = \mathcal{M}(P,g)$ is a smooth submanifold of dimension $d(P)$ as given in Equation 8 inside the space $\mathcal{B}_2^*(P)$.*

5.5. Orientability of $\mathcal{M}^*(P)$

We have now established that for a generic metric the moduli space is a smooth manifold. One other fact will be vital in the applications – that the moduli space is orientable.

Theorem 5.5.1. *Suppose that $\mathcal{M}^*(P) \subset \mathcal{B}_2^*(P)$ is a smooth submanifold in the sense that $H^2(\mathcal{E}(A)) = 0$ for every irreducible ASD connection A. Then $\mathcal{M}^*(P)$ is orientable. An orientation for this manifold is determined by choosing an orientation of $H^2_+(B;\mathbf{R})$ and $H^1(B;\mathbf{R})$.*

If M is a smooth n-manifold, then it is orientable if and only if the determinant line bundle $\Lambda^n TM$ of the tangent bundle (i.e., the top exterior power of the tangent bundle) is trivial. To orient M is to choose a trivialization of this bundle (see Problem 1). We construct a real line bundle over $\mathcal{B}_2^*(P)$ whose restriction to $\mathcal{M}^*(P)$ is the determinant line bundle of the tangent bundle of $\mathcal{M}^*(P)$. One then shows that the line bundle over $\mathcal{B}_2^*(P)$ is trivial.

The line bundle over $\mathcal{B}_2^*(P)$ is constructed as follows. Consider the family of elliptic operators parametrized by $\mathcal{A}_2^*(P)$

$$D_A = (\nabla_A^*, P_+\nabla_A) \colon \Omega^1_{L^2_2}(B;\mathrm{ad}P) \to \Omega^0_{L^2_1}(B;\mathrm{ad}P) \oplus \Omega^2_+(B;\mathrm{ad}P)_{L^2_1}.$$

We can form the determinant line bundle of this family of operators see [2]. It is a real line bundle over the parameter space $\mathcal{A}_2^*(P)$. The group $\mathcal{G}_3(P)$ acts linearly on this line bundle covering the natural action on $\mathcal{A}_2(P)$. Furthermore, $\{\pm 1\}$ acts trivially on the line bundle. Thus, there is an induced line bundle over the quotient space $\mathcal{B}_2^*(P)$. It is clear that over $\mathcal{M}^*(P)$ this line bundle agrees with the top exterior power of the tangent bundle of $\mathcal{M}^*(P)$.

Next, one must show that this line bundle is trivial. Since $\pi_1(\mathcal{B}^*(P))$ is either the trivial group or $\mathbf{Z}/2\mathbf{Z}$, one needs only to see that around at most one loop, the line bundle is trivial. There are simple geometric arguments to establish this fact, see [4].

Furthermore, since $\mathcal{B}_2^*(P)$ is connected, to orient the line bundle one needs only to orient the determinant line at one point of $\mathcal{B}_2^*(P)$. This we do at a connection which is trivial off of $c_2(P)$ small balls in B and in these balls is close to a standard charge-one ASD connection on the sphere. By an excision argument one shows that the determinant line at such a connection is isomorphic to the tensor product of the determinant line bundle for the trivial connection with $c_2(P)$ copies of the determinant line bundle for the charge-one instanton on the sphere. We choose once and for all an orientation for the determinant line bundle over space of charge one connections on the sphere. It remains to orient the determinant line bundle of the trivial bundle. This is accomplished (given an orientation for $\mathfrak{g}$) by orienting $H^2_+(B;\mathbf{R})$ and $H^1(B;\mathbf{R})$ (see Problem 14).

5.6. Variation of the metric

Since in the end we are looking for invariants of smooth manifolds not riemannian manifolds, we must understand what happens to the moduli space as we vary the metric.

Lemma 5.6.1. *The space $\mathcal{C}$ of riemannian metrics on B is a path-connected Banach manifold.*

Proof. We can view $\mathcal{C}$ as the space of sections of the bundle with fiber positive-definite symmetric matrices associated to the frame bundle of the tangent bundle of B and the action of $Gl(4,\mathbf{R})$ on 4×4-matrices by $(A,X)\mapsto AXA^{tr}$ for $A\in Gl(4,\mathbf{R})$ and X a 4×4 matrix. The set of positive-definite matrices is an open cone in a vector space, and it therefore has an affine structure, i.e., if X_0 and X_1 are positive definite matrices then so is $(1-t)X_0+tX_1$. This affine structure is invariant under the action of $Gl(4,\mathbf{R})$ and hence the set of cross sections of the fiber bundle has an affine structure. In particular, it is path-connected. Clearly, if $g_0, g_1\in\mathcal{C}$, i.e., if g_0 and g_1 are C^r riemannian metrics then the affine path between them is a smooth path in $\mathcal{C}$. □

This means that if we wish to compare two C^r-riemannian metrics g_0 and g_1 on B, we can joint them by a path.

Theorem 5.6.2. *Suppose that $g_0, g_1\in\mathcal{C}$. For any path $g_t, 0\le t\le 1$ we form the parametrized moduli space associated with the path*

$$\mathcal{M}^*(P,\{g_t\}) = \{([A],t)\}$$

where A is an irreducible g_t-ASD connection. This space sits naturally as a subspace in $\mathcal{B}^*_2(P)\times I$. If g_0, g_1 are in the G_δ subset described in Theorem 5.3.1, then for a generic smooth path $\{g_t\}$ joining them the parametrized moduli space $\mathcal{M}^*(P,\{g_t\})$ is a smooth, orientable submanifold of $\mathcal{B}^*_2(P)\times I$. A choice of orientation $\mathcal{O}$ for $H^2_+(B;\mathbf{R})$ and $H^1(B;\mathbf{R})$ determines an orientation for $\mathcal{M}^*(P,\{g_t\})$. Its oriented boundary is the disjoint union of the moduli space for g_1 with orientation induced by $\mathcal{O}$ and the moduli space for g_0 with the opposite orientation from the one induced by $\mathcal{O}$.

Proof. Since the projection mapping from $\mathcal{M}^*(P,\mathcal{C})\to\mathcal{C}$ is a Fredholm map, this result is a direct application of Sard's theorem. (Notice that we cannot necessarily arrange that the moduli spaces $\mathcal{M}^*(P,g_t)$ are smooth for every $t\in[0,1]$, only that the total space (the bordism) is smooth.) The orientability of $\mathcal{M}^*(P,\{g_t\})$ follows from the same argument that proves orientability of the moduli spaces. □

Theorem 5.6.3. *Suppose that $b_2^+(B)>1$ and that $g_0,g_1\in\mathcal{C}$ are generic riemannian metrics as in Theorem 5.4.3. Then for the generic smooth path $\{g_t\}\in\mathcal{C}$ connecting g_0 and g_1, we have $\mathcal{M}^*(P,\{g_t\})=\mathcal{M}(P,\{g_t\})$ (i.e., for every $t\in[0,1]$*

there are no reducible g_t-ASD connections) and $\mathcal{M}^*(P, \{g_t\})$ is a smooth orientable submanifold.

Proof. This is immediate from the previous theorem and Theorem 5.4.2. □

Remark 5.6.4. Notice that it is important here that $b_2^+(B) > 1$. When $b_2^+(B) = 1$, for the generic path there will be finitely many values of t for which there are reducible g_t-ASD connections. There are nice local models for these singular points in $\mathcal{M}(P, \{g_t\})$, but we do not have the time to go into this. (See [6] for more details.)

5.7. Problems

1. Let V be an n-dimensional real vector space. Compute $\dim \Lambda^i(V)$. Show that $\Lambda^n(V) \cong \mathbf{R}$, although the isomorphism is not canonical. Recall that an *orientation* of V is a choice of such an isomorphism, up to positive scalars.
2. Let V be an oriented real vector space with (positive-definite) inner product – a *euclidean* vector space. Let $SO(V)$ be the the space of orientation-preserving automorphisms of V preserving the inner product. Show that there is a natural group homomorphism
$$SO(V) \to SO(\Lambda^i(V)),$$
i.e. that the inner product on V naturally induces one on $\Lambda^i(V)$. If $\{e_i\}_{i=1}^n$ is an orthonormal basis for V, give a formula for the inner product $\langle e_{n_1} \wedge \dots \wedge e_{n_i}, e_{m_1} \wedge \dots \wedge e_{m_i}\rangle$.
3. Show that for a euclidean space V, wedge product induces a pairing
$$\Lambda^i(V) \otimes \Lambda^{n-i}(V) \to \Lambda^n(V) \cong \mathbf{R}$$
which is nondegenerate. Use this fact and the pairing canonically defined in Problem 2 to define a map
$$*\colon \Lambda^i(V) \to \Lambda^{n-i}(V).$$
What is $*(e_1 \wedge \cdots \wedge e_i)$? Show that $** = (-1)^{i(n-i)}$.
4. Use Problem 3 to define the Hodge $*$-operator
$$\Omega^i(B) \to \Omega^{n-i}(B)$$
for any riemannian manifold B. Indicate how to extend this to vector bundle valued differential forms.
5. Show that for a $4n$-dimensional vector space,
$$*\colon \Lambda^{2n}(V) \to \Lambda^{2n}(V)$$
is an orthogonal involution, i.e $*^2 = I$ and that $*$ preserves the natural inner product of $\Lambda^i(V)$. Show that is must consequently split Λ^{2n} into two orthogonal eigenspaces spaces $\Lambda_+(V)$ and $\Lambda_-(V)$, the $+1$ and -1 eigenspaces of $*$. Show that the two eigenspaces have the same dimension. Show that the natural map discussed in Problem 2 can be lifted to a map
$$SO(V) \to SO(\Lambda^+(V)) \times SO(\Lambda^-(V)).$$
Show that this map is injective.

6. Show that $\pi_1(SO(n)) \cong \mathbf{Z}/2\mathbf{Z}$ for $n > 2$. Use this to show that there is a simply-connected topological group, called $Spin(n)$ which forms a double-cover for $SO(n)$.
7. Identifying $SU(2)$ with the unit quaternions, show that conjugation on the "purely imaginary" quaternions induces a double cover of $SU(2) \to SO(3)$, i.e. $Spin(3) = SU(2)$.
8. Use the ideas from Problem 5 to show that the map
$$SO(4) \to SO(3) \times SO(3)$$
is a double cover. Use this result to show that
$$Spin(4) \cong Spin(3) \times Spin(3).$$
This phenomenon is unique to dimension 4 – all the other $Spin(n), n > 2$, are simple groups.
9. **A quaternionic approach to $Spin(4)$.** Let $\mathbf{H}$ denote the quaternion algebra. Consider the map
$$(SU(2) \times SU(2)) \times \mathbf{H} \to \mathbf{H}$$
given by
$$a \times b \times q \mapsto aqb^{-1}.$$
Show that this determines orthogonal action, preserving the norm on $\mathbf{H}$ induced by $q \mapsto q\bar{q}$, hence inducing a map
$$SU(2) \times SU(2) \to SO(4).$$
Show the map is surjective and identify its kernel. Compare these results with Problem 8.
10. Show that the operator $*$ on Λ^n of a $2n$-dimensional vector space V depends only on the conformal class of the inner product on V. Show that this fails in all other degrees.
11. Two riemannian metrics on a manifold are conformally equivalent if they differ by multiplication by a (real-valued) function. Show that the ASD condition depends only on the conformal class of the riemannian metric.
12. Show that if $F_0, F_1 \colon H_1 \to H_2$ are (linear) Fredholm maps of Hilbert spaces, and F_t is a homotopy between them through the space of Fredholm maps, then the index remains unchanged. Show that consequently, if $F \colon X \to Y$ is a map of connected Hilbert manifolds whose differential is always Fredholm, then $Ind(DF_x)$ at $x \in X$ is actually independent of a choice of point. What does this index say about the dimensions of preimages of points on Y?
13. Show that the symbol sequence in the proof of Lemma 5.2.5 is exact.
14. Let $\{F_t \colon H_1 \to H_2\}$ be a family of Fredholm operators depending smoothly on a parameter t in a parameter space T which is a Hilbert manifold. Define the determinant line bundle $\det(F) \to T$. Show that if $\mathcal{G}$ acts on $\det(F)$ linearly covering an action of $\mathcal{G}$ on T, and if for every $t \in T$ the subgroup $\mathrm{Stab}(t) \subset \mathcal{G}$ acts trivially on $\det(F_t)$, then there is a quotient line bundle $\det(F)/\mathcal{G} \to T/\mathcal{G}$.
15. Compute the determinant line associated to the product connection on the trivial bundle, and show that an orientation for the determinant line bundle is determined by orientations the real vector spaces $H^2_+(B;\mathbf{R})$ and $H^1(B;\mathbf{R})$.

16. Interpret the self-dual curvature map as a section of a vector bundle over $\mathcal{B}_2^*(P)$. Show that this section has a Fredholm differential at any point mapping to 0. Show that the preimage of zero under this section is the moduli space.
17. Show that the formula for the Euler characteristic of the basic elliptic complex $\mathcal{E}(A)$ is correct in the case when P is a trivial $SU(2)$-bundle.
18. Work out the corresponding formula for the Euler characteristic of the basic elliptic complex in the case of $SO(3)$-bundles.
19. Let B be a compact, oriented 4-manifold. Show that $b_0(B)-b_1(B)+b_2^+(B) = (\sigma(B)+\chi(B))/2$ where $\sigma(B)$ is the signature of B and $\chi(B)$ is the Euler characteristic of B.

Three-dimensional motivation for the ASD condition.

20. Let (N, g_N) be a riemannian three-manifold. Consider the four-manifold $B = N \times \mathbf{R}$, with the metric $g_B = dt^2 + g_N$. Show that

$$\Omega^2(B) \cong p_N^*(\Omega^1(N)) \oplus p_N^*(\Omega^2(N)).$$

Show that with respect to this decomposition, we can write $*_B = *_N \oplus *_N$.

21. Note that a path of connections A_t over N can be viewed as a map

$$a\colon \mathbf{R} \to \mathcal{A}(N \times SU(2)),$$

where

$$A_t = \nabla + a(t)$$

with ∇ the product connection. In turn, this determines a connection over B by

$$\frac{\partial}{\partial t} + \nabla + a.$$

Using the previous exercise, show that the path A_t satisfies the gradient flow equation for CS (see Problem 24 Section IV) iff the corresponding connection over B is ASD.

Yang-Mills functional approach to the ASD condition.

22. Suppose B is a riemannian manifold, $\omega \in \Omega^2(B)$. Show that for the induced metric on Ω^2,

$$\omega \wedge \omega = \langle \omega^+, \omega^+ \rangle - \langle \omega^-, \omega^- \rangle$$

Use this to show that for $A \in \mathcal{A}(P)$,

$$8\pi^2 \int_B c_2(B) = \int_B Tr(F_A \wedge F_A) = 2\left(\|P_- F_A\|_{L^2}^2 - \|P_+ F_A\|_{L^2}^2\right).$$

23. Consider the *Yang-Mills functional* on the space of connections over $P \to B$

$$\mathcal{YM}\colon \mathcal{A} \to \mathbf{R}$$

defined by

$$\mathcal{YM}(A) = \|F_A\|_{L^2}^2.$$

Show that this map factors through $\mathcal{B}_2(P)$. The functional $\mathcal{YM}(A)$ is often called the *Yang-Mills energy*, or simply the energy of A.

24. Show that the global minima for the Yang-Mills functional are precisely the ASD connections. Evaluate the minimum value of $\mathcal{YM}$ if it exists.
25. Show all ASD connections over the trivial bundle $B \times SU(2)$ are trivial.

Complex approach to the ASD condition

26. Let V be a two-dimensional complex vector space. Show that $\Lambda^2_{\mathbf{R}}(V) \otimes_{\mathbf{R}} \mathbf{C}$ splits naturally into $\Lambda^{2,0}(V) \oplus \Lambda^{0,2}(V) \oplus \Lambda^{1,1}(V)$, a pair of one dimensional (complex) vector spaces and a four-dimensional one. By naturality here, we mean that the action of $Gl(V)$ preserves this splitting.
27. Show that a hermitian form h on V naturally distinguishes a nonzero element ω of $\Lambda^{1,1}(V)$ by the formula
$$\omega(X,Y) = \mathrm{Im}\,(h(X,Y))\,.$$
28. Show that
$$\Lambda^2_-(V) \otimes_{\mathbf{R}} \mathbf{C} = \Lambda^{2,0}(V) \oplus \Lambda^{0,2}(V) \oplus \mathbf{C}\omega$$
and
$$\Lambda^2_+(V) \otimes_{\mathbf{R}} \mathbf{C} = \omega^\perp \cap \Lambda^{1,1}(V),$$
where $\Lambda^2_+(V)$, and $\Lambda^2_-(V)$ are taken with respect to the natural euclidean structure on V (see Problem 5). Show more generally, that given a complex euclidean vector space V, a choice of hermitian structure is equivalent to a choice of ASD 2-form. This is why one studies anti-self-dual, rather than self-dual connections.
29. Let B be a closed, oriented, 4-manifold and let $P \to B$ be a principal $SU(2)$-bundle. Given an element $\sigma \in \mathcal{G}_3(P)$ we construct a principal bundle over $B\times S^1$ where the total space is given by $P\times I$ modulo the equivalence relation $(p,1) \cong (\sigma(p),0)$ Show that this sets up a bijective correspondence between $\pi_0(\mathcal{G}_3(P))$ and the set of isomorphism classes of principal $SU(2)$-bundles over $B \times S^1$ which are isomorphic to P when restricted to $B \times \{0\}$.
30. Show that if B is simply connected then there are at most two isomorphism classes of principal $SU(2)$-bundles over $B \times S^1$ which are isomorphic to P when restricted to $B \times \{0\}$. Show that these two bundles are isomorphic over the complement of a point of $B \times S^1$.
31. Show that under the conditions of Problem 30 there are two distinct principal $SU(2)$-bundles over $B\times S^1$ isomorphic to P over $B\times\{0\}$ if and only if B is a spin 4-manifold. Hint: Given a principal $SU(2)$-bundle Q over $B\times S^1$ we can vary it over the top cell by the non-trivial element in $\pi_5(BSU(2)) = \mathbf{Z}/2\mathbf{Z}$ to produce a new bundle $Q\#\eta$. These bundles are isomorphic if and only if all bundles of the type we are considering over $B \times S^1$ are isomorphic. The first k-invariant of $BSU(2)$ is an element in $H^6(K(\mathbf{Z},4);\mathbf{Z}/2\mathbf{Z})$. This group is equal to $\mathbf{Z}/2\mathbf{Z}$ and the non-trivial element is $Sq^2(\iota_4)$ where Sq^2 is the second Steenrod square and $\iota_4 \in H^4(K(\mathbf{Z},4);\mathbf{Z}/2\mathbf{Z})$ is the reduction modulo two of the fundamental class. Using this, show that we can produce an isomorphism from Q to $Q\#\eta$ if and only if there is a class $x \in H^2(B;\mathbf{Z})$ with the property that $Sq^2(x) \neq 0$.
32. Prove that if B is a closed, simply connected 4-manifold then $\pi_0(\mathcal{G}(P))$ is trivial if B is not a spin manifold and is isomorphic to $\mathbf{Z}/2\mathbf{Z}$ if B is a spin manifold.
33. Show that with B as in the previous problem and spin, the element $-1 \in \mathcal{G}$ is in the non-identity component of of $\mathcal{G}(P)$ if and only if $c_2(P)$ is odd.

34. Show that if B is a closed, simply connected 4-manifold then $\pi_1(\mathcal{B}_2^*(P))$ is trivial unless B is a spin manifold and $c_2(P)$ is even. In this case $\pi_1(\mathcal{B}_2^*(P)) = \mathbf{Z}/2\mathbf{Z}$.
35. Let $H \subset G$ be a closed subgroup. Let Q be a principal H-bundle. Show that connections on Q induce connections on the principal G-bundle $P = Q \times_H G$. Show that a connection on $SU(2)$-bundle is reducible iff it arises in this way from a connection on a principal S^1-bundle. Moreover, show that the reducible $SU(2)$-connection is ASD iff the related S^1 connection is.
36. Let $Q \to B$ be a principal S^1-bundle, and let h be a closed two-form $h \in \Omega^2(B; \mathbf{R})$, representing 2π times the first Chern class of Q. Show that h is the curvature form for some connection on Q. (Note that there is a canonical identification of adQ with $B \times \mathbf{R}$.)
37. Prove Lemma 5.2.4.
38. Derive a formula for the L^2 adjoint of d, d^*, in terms of the Hodge-star $*$ and d. Use this formula to show that any closed, ASD 2-form is harmonic.

LECTURE 6
Compactness and Gluing Theorems

Uhlenbeck compactness and the Taubes' gluing theorem

We continue with the notation and assumptions of the previous section: B is a closed, smooth, oriented 4-manifold and $P \to B$ is a non-trivial principal $SU(2)$-bundle.

In the last section we established several nice properties of the moduli space of g-ASD connections assuming that g is a generic riemannian metric on B. Before we can make applications to the study of the smooth topology of 4-manifolds we need one more ingredient – compactness results. Unfortunately, the moduli spaces are not usually compact. Fortunately, because of theorems of Uhlenbeck and Taubes we have good control on the nature of neighborhoods of infinity. These results are sufficiently strong that they allow us to define compactifications of the moduli spaces which will meet our needs.

The construction of these compactifications and the study of their basic properties comprise the bulk of this section. At the end of the section we give a first application of the compactified moduli space – in this application we study simply connected 4-manifolds with negative definite intersection forms. Here we are following history, for this was Donaldson's first result, the first result about 4-manifolds proved using gauge theory. It is the result that introduced gauge theory to topologists.

6.1. Uhlenbeck compactness

The Uhlenbeck compactness theorem tells us what must be happening if we have a sequence of ASD connections without a convergent subsequence.

Theorem 6.1.1. *Let B be a closed oriented riemannian 4-manifold and let $P \to B$ be a principal $SU(2)$-bundle with $c_2(P) = k > 0$. Let $\{A_n\}$ be a sequence of ASD L^2_2-connections on P. Then after passing to a subsequence we can find:*

- *a principal $SU(2)$-bundle $P' \to B$ with $0 \leq c_2(P') = k' \leq k$,*
- *points $\{x_1, \ldots, x_t\}$ in B*
- *positive integers $\{m_1, \ldots, m_t\}$*

- *an ASD connection A' on P', and*
- *bundle isomorphisms $\sigma_n\colon P'|_{B-\{x_1,\dots,x_t\}} \to P|_{B-\{x_1,\dots,x_t\}}$ which are L^2_3 over every compact subset $K \subset B - \{x_1, \dots, x_t\}$,*

such that:

- *For any compact subset $K \subset B - \{x_1, \dots, x_k\}$ the connections $\sigma_n^*(A_n|_K)$ converge in $L^2_2(K)$ to $A'|_K$.*
- *The L^1-functions $|F_{A_n}|^2$ on B converge in the weak topology to the measure*

$$|F_{A'}|^2 + \sum_{i=1}^{t} 8\pi^2 m_i \delta_{x_i},$$

where δ_{x_i} denotes the δ-mass at x_i of mass 1.

Remark 6.1.2. If follows from the last condition that $c_2(P) = c_2(P') + \sum_{i=1}^t m_i$. Since $c_2(P') \geq 0$ and since the $m_i > 0$ it follows that $t \leq k$.

There are two local results, due to Uhlenbeck, which are used in the proof of this theorem. Let $V \subset B$ be a geodesic ball of radius $r > 0$ sufficiently small. Let $V' \subset V$ be a concentric ball of smaller radius. The first result says that there are universal constants $\epsilon > 0$ and C such that if A is any L^2_2-connection on a principal bundle P with a C^∞ product connection A_0 over V with $\|F_A\|_{L^2(V)} \leq \epsilon$ then there is an L^2_3-gauge transformation σ on P such that

$$\|\sigma^* A - A_0\|_{L^2_1} \leq C\|F_A\|_{L^2}$$

and

$$\nabla^*_{A_0}(\sigma^* A - A_0) = 0.$$

Furthermore, if A is an ASD L^2_2-connection on $P \to V$ then $\sigma^* A$ is C^∞ and we have the estimate

$$\|\sigma^* A - A_0\|_{L^2_3(V')} \leq C\|F_A\|_{L^2(V)}.$$

It follows from this result that if A_n is a sequence of ASD L^2_2-connections on a small ball V in B and if the L^2-norms of the curvatures of the A_n are all less than ϵ, then for any smaller ball $V' \subset V$ after passing to a subsequence there is a sequence of L^2_3-changes of gauge $\{\sigma_n\}$ such that the sequence $\sigma_n^* A_n$ converges over V' in L^2_2 to a limiting connection.

The second result is the removable singularities result. It says if A is an ASD connection on a principal bundle Q_0 over a punctured ball in B which is L^2_2 over any compact subset of the punctured ball, and if the L^2-norm of the curvature of A is finite, then there is an extension of Q_0 to a principal bundle Q over the entire ball and an extension of A to an ASD L^2_2-connection on Q.

Proofs of these results can be found in [**4**] and [**3**].

We fix $\epsilon > 0$ as in the first of Uhlenbeck's results above. Now we consider a sequence A_n of ASD L^2_2-connections on P. We have the sequence of L^1-functions $|F_{A_n}|^2$ on B of constant L^1-norm. After passing to a subsequence, we can assume that these functions converge weakly to a measure λ. Let us consider the set of points $\mathcal{S}$ of all $x \in B$ for which λ has a δ-mass of norm at least $\epsilon/2$. Since the norm of λ is $8\pi^2 c_2(P)$, there are only finitely many such points.

Let us first consider any point $x \notin \mathcal{S}$. According to the first result cited above, after passing to a subsequence we can find the following. There is a small ball V' about x and L^2_3-changes of gauge σ_n for $P|_{V'}$ so that $\sigma_n^* A_n$ converges in $L^2_2(V')$

to an ASD connection $A_\infty(V')$ on $P|_{V'}$. We can cover $B - \mathcal{S}$ by countably many such balls V and take a diagonal subsequence, so that the above statement holds for this subsequence and all points of $B - \mathcal{S}$.

Now a local version of Lemma 4.4.6 shows that after passing to a further subsequence we can assume that the local changes of gauge glue together to form global changes of gauge σ_n on $B - \mathcal{S}$ such that for any compact subset $K \subset B - \mathcal{S}$ the σ_n lie in $L^2_3(K)$ and such that $\sigma_n^* A_n$ converges in $L^2_2(K)$ to an ASD $A'_0|_K$. Said another way, we have constructed a principal bundle P'_0 over $B - \mathcal{S}$ and bundle isomorphisms

$$\sigma_n \colon P'_0 \to P|_{B-\mathcal{S}}$$

and an ASD connection A'_0 for P'_0 such that $\sigma_n^* A_n$ converge in L^2_2 on any compact set to A'_0. The σ_n are contained in $L^2_3(K)$ for any compact subset $K \subset B - \mathcal{S}$. On each compact subset $K \subset B - \mathcal{S}$, the local L^1-function of $|F_{A'_0}|^2$ is the limit of $|F_{A_n}|^2$. Hence the L^1-norm of $|F_{A'_0}|^2$ on any compact subset of $B - \mathcal{S}$ is at most $8\pi^2 c_2(P)$. This proves that the $\|F_{A'_0}\|^2_{L^2(B-S)}$ is finite, and in fact bounded above by $8\pi^2 c_2(P)$. Now by the removable singularities theorem, we can extend P'_0 to a principal bundle $P' \to B$ in such a way that A'_0 extends to an ASD connection A' on P'. We know that

$$c_2(P') = \frac{1}{8\pi^2} \int_B |F_{A'}|^2 d\mathrm{Vol}$$

and thus we see that $0 \leq c_2(P') \leq c_2(P)$.

The last thing that we need to show is that at any point $x \in \mathcal{S}$ the δ-mass of λ at x is of the form $8\pi^2 m$ for some positive integer m. Since by the definition of $\mathcal{S}$ we know that this δ-mass is positive, we need only show that it is of the form $8\pi^2 m$ for an integer m. Take a small ball V in B centered at x, small enough so that its closure is disjoint from all other points of $\mathcal{S}$. Consider two connections – the restriction of A_n to $P|_V$ and the restriction of A' to $P'|_V$. We use σ_n to identify these bundles in a collar neighborhood of ∂V. For $n >> 0$ the L^2_2-distance between the restrictions of these connections to the collar C of ∂V is close to zero. Thus, for all $n >> 0$ we can construct a connection on $P \cup P' \to V \cup -V$ by identifying the copies of C in V by the identity and the bundles over them by σ_n. On this principal bundle there is an L^2_2 connection which is close to A_n on $P|_V$ and close to A' on P' in the L^2_2-topology.

The base $V \cup -V$ is a closed manifold. Hence, the second Chern form of the connection over $V \cup -V$ has integral equal to some integer m. That means that the sum

$$\int_V |F_{A_n}|^2 - \int_V |F_{A'}|^2 + \int_C |F_{A'}|^2$$

is approximately equal to $8\pi^2 m$. In fact as $n \mapsto \infty$ the limiting value of these quantities is $8\pi^2 m$. This is true for any small ball V about x. Since A' is a connection on all of $P'|_V$, we see that

$$\lim_{V \mapsto \{x\}} \int_V |F_{A'}|^2 = 0.$$

Thus, we have

$$\lim_{V \mapsto \{x\}} \lim_{n \mapsto \infty} \int_V |F_{A_n}|^2 = 8\pi^2 m.$$

This is exactly the statement that the weak limit λ of the functions $|F_{A_n}|^2$ has a δ-mass at x of mass $8\pi^2 m$.

6.2. Gluing together connections

The Uhlenbeck compactness theorem gives us partial control on the nature of infinity, but the full analysis that is required in the applications requires an explicit description of neighborhoods of infinity. This analysis is supplied by Taubes' gluing theorems. Before stating these theorems, we need to make some preliminary remarks about connected sums which will permit us to glue together connections in a way which is consistent with the connected sum of manifolds.

Suppose that B and B' are oriented riemannian 4-manifolds and $V_\lambda(x) \subset B$ and $V_\lambda(x') \subset B'$ are geodesic balls of radius $0 < \lambda < 1$ centered at x and x' respectively. Suppose that we have an orientation-reversing isometry $\tau\colon TB_x \to TB'_{x'}$. Using this isometry and exponentiation, we define an orientation-reversing diffeomorphism from $V_\lambda(x) \to V_\lambda(x')$ whose differential at x is τ. We identify these balls by this diffeomorphism. Now let $\iota_\lambda\colon V_\lambda(x) \mapsto V_{\lambda^3}(x')$ be the usual inversion in the sphere of radius λ^2 centered at x. That is to say

$$\iota_\lambda(\xi) = \frac{\lambda^4 \tau(\xi)}{|\xi|^2}$$

with respect to the local geodesic normal coordinates centered at x and x'.

The resulting manifold

$$B\#_\lambda B' = (B - V_{\lambda^3}(x)) \cup_{\iota_\lambda} (B' - V_{\lambda^3}(x'))$$

is a representative for the connected sum.

If the riemannian metrics on B and B' are standard on $V_\lambda(x)$ and $V_\lambda(x')$ then ι_λ is a conformal isomorphism. If the riemannian metrics on B and B' are nearly standard on $V_\lambda(x)$ and $V_\lambda(x')$ (in the sense that if we scale these balls up to unit size then the metrics are almost standard), then ι_λ is a nearly a conformal isomorphism. In this case by patching the metrics on B and B' together by a partition of unity we construct a metric on this representative for the connected sum which agrees with the original metrics on $B - V_\lambda(x)$ and $B' - V_\lambda(x')$, and is nearly conformally equivalent to these metrics even on the annulus $V_\lambda(x) - V_{\lambda^3}(x)$. Henceforth, we will implicitly use such a metric.

Now suppose that $P \to B$ and $P' \to B'$ are principal bundles with the same structure group, and suppose that $P|_{V_\lambda(x)}$ and $P'|_{V_\lambda(x')}$ have product structures. Then given a G-equivariant isomorphism $\sigma\colon P_x \to P'_{x'}$ we use the trivializations to give an identification of $P|_{V_\lambda(x)}$ with $P'|_{V_\lambda(x')}$. With this identification we can then glue P and P' together to form a principal bundle $P\#_\lambda P' \to B\#_\lambda B'$. The isomorphism class of this principal bundle is independent of the choice of trivializations (see Problem 20).

Now suppose in addition that A and A' are connections on P and P' which are flat on $V_\lambda(x)$ and $V_\lambda(x')$ respectively. Use these to define product structures for $P|_{V_\lambda(x)}$ and $P'|_{V_\lambda(x')}$, allowing us to form $P\#_\lambda P'$. Then any $\sigma\colon P_x \to P_{x'}$ produces a glued-up bundle $P\#_\lambda P'$ on which there is a glued-up connection $A\#_\lambda A'$. (See Problem 20).

We need a more general construction which approximates the above. Suppose now that we have balls $V_r(x), V_r(x')$ for $r > \lambda$ and trivializations of P and P'

over these balls with respect to which the connections A and A' have small L^2_1-norm. Then using bump functions whose derivatives are supported in the annuli $V_r(x) - V_\lambda(x)$ and $V_r(x') - V_\lambda(x')$ we can damp A and A' to produce connections $\tilde{A}$ and $\tilde{A}'$ which agree with A and A' on $B - V_r(x)$ and $B' - V_r(x')$ respectively and which are product connections in the given product structures for P and P' over $V_\lambda(x)$ and $V_\lambda(x')$. It is an easy exercise using the Sobolev multiplication theorems to show that there is a constant C depending only on r and λ, which is bounded for fixed r as $\lambda \mapsto 0$, so that

$$\int_W |F_{\tilde{A}\#_\lambda \tilde{A}'}|^2 \leq C\left(\|A\|^2_{L^2_1(V_r(x))} + \|A'\|^2_{L^2_1(V_r(x'))}\right),$$

where

$$W = (V_r(x) - V_{\lambda^3}(x)) \cup (V_r(x') - V_{\lambda^3}(x')) \subset B\#_\lambda B'.$$

This gives the basic geometric construction underlying the Taubes' gluing theorem.

6.3. Taubes' gluing theorem

Before we can glue together ASD connections we need a slightly stronger version of the first result of Uhlenbeck's cited earlier in this section. Fix $r > 0$ small enough that the metric of B is nearly standard on the ball $V_{2r}(x, X) \subset B$ for some $x \in B$ and likewise for $x' \in B'$.

Lemma 6.3.1. *Suppose that A is an ASD L^2_2-connection on a principal bundle P over the ball $V_{2r}(x) \subset B$. Then there is a universal constant $\epsilon > 0$ and a constant C depending only on r such that if*

$$\|F_A\|_{L^2(V_{2r}(x))} < \epsilon$$

then there is a flat L^2_2 connection A_0 on P (inducing a product structure) such that

$$\|A - A_0\|_{L^2_2(V_r(x))} \leq C\|F_A\|_{L^2(V_{2r}(x))}.$$

In fact, there is a canonical choice of A_0 which varies smoothly with A. The resulting product structure is called the Coulomb gauge *of A.*

We are now in position to glue ASD connections together to get almost ASD connections over the connected sum.

Corollary 6.3.2. *Given $\epsilon > 0$ there is $\delta > 0$ such that the following holds. Fix $P \to B$ and $P' \to B'$ principal bundles; fix points $x \in B$ and $x' \in B'$, and fix isomorphisms $\sigma\colon P_x \to P'_{x'}$ and $\tau\colon TB_x \to TB'_{x'}$ as before. Finally, fix $r > 0$ sufficiently small. Let A and A' be ASD L^2_2-connections on P and P'. Suppose that*

$$\|F_A\|_{L^2(V_{2r}(x))}, \|F_{A'}\|_{L^2(V_{2r}(x'))} < \delta,$$

then for all $\lambda > 0$ sufficiently small there is a connection $\tilde{A}\#_\lambda\tilde{A}'$ on $P\#_\lambda P'$ obtained by using the Coulomb gauges for A and A' over $V_r(x)$ and $V_r(x')$ to damp out to connections $\tilde{A}$ and $\tilde{A}'$ flat on $V_\lambda(x)$ and $V_\lambda(x')$, respectively, then and using the identification σ to glue these new connections together. The resulting connection satisfies the following properties:

- *$\tilde{A}\#_\lambda\tilde{A}'$ varies smoothly with A, A',*
- *$\tilde{A}\#_\lambda\tilde{A}'$ is flat on $(V_\lambda(x) - V_{\lambda^3}(x)) \subset B\#_\lambda B'$,*
- *$\tilde{A}\#_\lambda\tilde{A}'$ agrees with A on $P|_{B-V_r(x)}$ and with A' on $P'|_{B'-V_r(x')}$, and*

- $\|P_+\left(F_{\tilde{A}\#_\lambda\tilde{A}'}\right)\|_{L^2_1} < \epsilon$

The above construction can be done smoothly in families, as follows. Fix $V_{2r}(x) \subset B$ and $V_{2r}(x') \subset B'$ as above. Define $\widetilde{\mathcal{U}}(\delta) \subset \widetilde{\mathcal{M}}(P)$ to be the set of ASD connections A for which

$$\|F_A\|_{L^2(V_{2r}(x))} < \delta,$$

and define $\widetilde{\mathcal{U}}'(\delta) \subset \widetilde{\mathcal{M}}(P')$ similarly. The quotients of the change of gauge actions on these subsets are denoted $\mathcal{U}(\delta)$ and $\mathcal{U}'(\delta)$.

Corollary 6.3.3. *For any fixed $\epsilon > 0$, there is a corresponding choice of $\delta > 0$, $\lambda > 0$, and $r > \lambda$ with respect to which the construction outlined in the previous corollary defines a smooth map*

$$\mathrm{ISO}_{SU(2)}(P_x, P'_{x'}) \times \widetilde{\mathcal{U}}(\delta) \times \widetilde{\mathcal{U}}'(\delta) \to \mathcal{A}_2(P\#_\lambda P').$$

Here $\mathrm{ISO}_{SU(2)}(P_x, P'_{x'})$ is the space of $SU(2)$-equivariant isomorphism from P_x to $P'_{x'}$. (It is non-canonically diffeomorphic to $SU(2)$.) This map is equivariant with respect to changes of gauge (see Problem 22) and hence induces a smooth map

$$\left(\mathrm{ISO}_{SU(2)}(P_x, P'_{x'}) \times \widetilde{\mathcal{U}}(\delta) \times \widetilde{\mathcal{U}}'(\delta)\right) / \left(\mathcal{G}_3(P) \times \mathcal{G}_3(P')\right) \to \mathcal{B}_2(P\#_\lambda P').$$

Now the heart of the Taubes gluing argument is to prove that under favorable circumstances one can canonically deform the glued up connection $\tilde{A}\#_\lambda\tilde{A}'$ to an ASD connection over $B\#_\lambda B'$.

Theorem 6.3.4. *Fix $P \to B$, $P' \to B'$, $x \in B, x' \in B'$. Also fix σ, τ as above. Let A, A' be ASD connections at which the spaces $\widetilde{\mathcal{M}}(P)$ and $\widetilde{\mathcal{M}}(P')$ are smooth. Then for all sufficiently small choices for r and λ (depending on A and A'), there is a unique point in $\widetilde{\mathcal{M}}(P\#_\lambda P')$ minimizing the L^2_2-distance to $\tilde{A}\#_\lambda\tilde{A}'$. This closest point varies smoothly in families in the following sense. Given A, A' there are $\mathcal{G}_3(P)$- and $\mathcal{G}_3(P')$-invariant neighborhoods $\widetilde{\mathcal{U}}, \widetilde{\mathcal{U}}'$ of A, A' in $\widetilde{\mathcal{M}}(P), \widetilde{\mathcal{M}}(P')$ such that for all sufficiently small r,λ, and any τ, there is a smooth map*

$$T\colon \mathrm{ISO}_{SU(2)}(P_x, P'_{x'}) \times \widetilde{\mathcal{U}} \times \widetilde{\mathcal{U}}' \to \widetilde{\mathcal{M}}(P\#_\lambda P').$$

This map is characterized by the property that $T(\sigma, A, A')$ is the point in $\widetilde{\mathcal{M}}(P\#P')$ with minimal L^2_2-distance to $A\#_\lambda A'$. Moreover, the image of this map contains all connections in $\widetilde{\mathcal{M}}(P\#_\lambda P')$ whose restriction to $B \setminus V_r(x)$, resp. $B' \setminus V_r(x)$, is L^2_2-close to the restriction of A, resp. A'. The map T is equivariant and induces a diffeomorphism onto an open subset

$$T\colon \left(\mathrm{ISO}_{SU(2)}(P_x, P'_{x'}) \times \widetilde{\mathcal{U}} \times \widetilde{\mathcal{U}}'\right) / \left(\mathcal{G}_3(P) \times \mathcal{G}_3(P')\right) \to \mathcal{M}(P\#_\lambda P').$$

Remark 6.3.5. Note that the space of gluing parameters

$$\left(\mathrm{ISO}_{SU(2)}(P_x, P'_{x'}) \times \widetilde{\mathcal{U}} \times \widetilde{\mathcal{U}}'\right) / \left(\mathcal{G}_3(P) \times \mathcal{G}_3(P')\right)$$

is a finite-dimensional object. In fact, letting $\mathcal{U}, \mathcal{U}'$ denote the gauge group quotients of $\widetilde{\mathcal{U}}, \widetilde{\mathcal{U}}'$, the parameter space maps to $\mathcal{U} \times \mathcal{U}' \subset \mathcal{M}(P) \times \mathcal{M}(P')$ with each fiber having dimension at most 3.

Remark 6.3.6. Taubes' result can be formulated in the case where the points are not smooth (See [**9**]), but we have left out the statement for the sake of exposition.

6.4. The moduli space of instantons over S^4

We wish to use Taubes' Gluing Theorem to study the ends of $\mathcal{M}(P)$. The idea is that a connection whose curvature measure is nearly singular at a point x can be obtained by gluing together (via the above construction) a lower charge ASD connection over B and an ASD connection over S^4. From this point of view, it becomes clear that connections over S^4 are fundamental objects of study. For this reason, we turn our attention to these, first concentrating on the charge-one case.

Let $Q_1 \to S^4$ be the principal bundle over S^4 with $c_2(Q_1) = 1$ and give S^4 the round metric. Consider the action of $\text{Diff}(S^4)$ on $\mathcal{B}(Q_1)$. The conformal invariance of the ASD condition ensures that the ASD moduli space $\mathcal{M}(Q_1)$ is preserved by the action of the subgroup $\text{Conf}(S^4) \subset \text{Diff}(S^4)$ of conformal maps of S^4 (see Problem 15). In turn, this restricted action allows us to gain an explicit understanding of the moduli space $\mathcal{M}(Q_1)$ when $c_2(Q_1) = 1$, as we will outline presently.

Any positive measure ν with total mass 1 on S^4 has a well-defined center of mass in $\overline{D}^5$, lying in the open 5-ball D^5, unless ν is a delta-mass concentrated at a single point in S^4. In particular, any ASD connection on Q_1 has a center of mass in D^5 defined to be the center of mass of the measure

$$\frac{|F_A|^2 d\text{Vol}}{8\pi^2}.$$

Given any nonsingular, mass 1 measure ν, the map

$$\text{Conf}(S^4) \to D^5$$

taking a conformal transformation ϕ to the center of mass of $\phi^*(\nu)$ induces an identification

$$\text{Conf}(S^4)/SO(4) \cong D^5.$$

It follows that the orbit under $\text{Conf}(S^4)$ of any connection is at least five-dimensional. In fact, the orbit of a generic connection in $\mathcal{B}(Q_1)$ will be eleven-dimensional – it will have no stabilizers. However, the ASD connections are far from generic:

Lemma 6.4.1. *For any* $[A] \in \mathcal{M}(Q_1)$, *the stabilizer of* A *in* $\text{Conf}(S^4)$ *is isomorphic to* $SO(4)$.

Proof. The stabilizer of a connection A must stabilize its associated curvature measure, so it must be conjugate to a subgroup of $SO(4)$. It cannot be conjugate to a proper subgroup of $SO(4)$, because if it were, the dimension of orbit of A would have to exceed 5, but we know from the Atiyah-Singer index calculation from Lecture V that $dim(\mathcal{M}(A)) = 5$. Technically, in order to invoke the Atiyah-Singer index theorem, one must know that $\mathcal{M}(Q_1)$ is smooth. But this is true because S^4 has a metric of positive curvature (See Problem 8).) □

Since the action of $\text{Conf}(S^4)$ on the moduli space is proper (see Problem 16), it follows from the lemmas we have outlined that the moduli space must decompose as a disjoint union of five-dimensional balls, the orbits of the action of $\text{Conf}(S^4)$. By Uhlenbeck's compactness theorem there can only be finitely many five balls in this decomposition. One can show that the ASD moduli space on S^4 is nonempty (see Problem 12), by giving an explicit connection matrix. It is a deep result of Atiyah [**1**] that the moduli space consists of only one five-ball.

Theorem 6.4.2. *The moduli space* $\mathcal{M}(Q_1)$ *is the five-dimensional ball, which is isomorphic as a* $\mathrm{Conf}(S^4)$*-space to* $\mathrm{Conf}(S^4)/SO(4)$*. Thus, the gauge equivalence class of a connection is uniquely determined by the center of mass in* D^5 *of its curvature measure.*

A connection is said to be *centered* at a point $x \in S^4$ if its center of mass is closer to x than it is to any other point in ∂D.

This definition can be made for any principal bundle over S^4 of higher Chern class, defining the center of mass for a connection A to be the center of mass in D^5 of the measure

$$\frac{|F_A|^2 d\mathrm{Vol}}{8\pi^2 c_2}.$$

We do not have as much explicit information about the moduli spaces for bundles of higher Chern class on S^4 as we do about that for the bundle Q_1. Nevertheless, there is one important qualitative result which is absolutely essential: for any bundle $P \to S^4$ the moduli space of ASD connections on P (with respect to the round metric on S^4) are smooth (see Problem 8).

6.5. The ends of $\mathcal{M}(P)$

The Taubes' Gluing Theorem can be modified to parameterize neighborhoods of infinity of the moduli space $\mathcal{M}(P)$. To do this, we use the identification $B\#S^4 \cong B$. More precisely, letting $s \in S^4$ denote the south pole on S^4 with the round metric and any point $x \in B$, for sufficiently small λ, $B\#_\lambda S^4$ is diffeomorphic to B under a diffeomorphism which is an isometry away from a neighborhood of x and which is close to a conformal isometry near x, how close depending on λ.

To get a reasonable understanding of a neighborhood of infinity of $\mathcal{M}(P)$, we must modify the gluing parameters a bit. Rather than choosing a fixed connected sum point on B, we will find it convenient to allow the point to vary. This alone would increase the dimension of the parameter space. To balance this effect, we restrict to certain class of connections over S^4, the ones which are concentrated at the point antipodal to the gluing point. Explicitly, fix a point $s \in S^4$, and let $Q \to S^4$ denote the principal $SU(2)$ bundle over S^4 with $c_2(Q) = t > 0$. Letting $n \in S^4$ denote the point antipodal to s, let $\widetilde{Z}_n(Q,\delta)$ denote the space of ASD connections over S^4 which are centered over the point n and whose energy over $V_{2r}(s)$ is bounded above by δ.

Let $P' \to B$ be a principal $SU(2)$-bundle such that $c_2(P') + c_2(Q) = c_2(P)$. Let $N = \mathrm{ISO}_{SU(2)}(P', Q_s)$ denote the locally trivial fiber bundle whose fiber over $x \in B$ is $\mathrm{ISO}_{SU(2)}(P'_x, Q_s)$. Also denote by $F = \mathrm{ISO}^-_{SO(4)}(F(B), F_s(S^4))$ the locally trivial fiber bundle of orientation-reversing orthogonal isomorphisms between the frame bundle of the tangent bundle of B and the space of frames at s for the tangent bundle of S^4. For any subset $\mathcal{U} \subset \mathcal{M}(P')$, we denote by $\widetilde{\mathcal{U}} \subset \widetilde{\mathcal{M}}(P')$ the preimage of $\mathcal{U}$.

Theorem 6.5.1. *For any pre-compact, open subset* $\mathcal{U} \subset \mathcal{M}(P')$ *for which the lift* $\widetilde{\mathcal{U}} \subset \widetilde{\mathcal{M}}(P')$ *is contained in the subset of smooth points, there is a gluing map*

$$T\colon \widetilde{\mathcal{U}} \times \widetilde{Z}_n(Q,\delta) \times (F \times_B N) \,/\, (\mathcal{G}_3(P') \times \mathcal{G}_3(Q) \times SO(4)) \to \mathcal{M}(P),$$

which is a diffeomorphism onto an open subset. The image of this map consists of points in $\mathcal{M}(P)$ which, when restricted to the complement of some geodesic ball of sufficiently small radius, are L^2_2-close, up to change of gauge, to a connection representing a point in $\mathcal{U}$.

Remark 6.5.2. Let us describe the action of the automorphism group on the gluing data. The group $\mathcal{G}_3(P')$ acts on $\widetilde{\mathcal{U}}$ and N in the natural way, the group $\mathcal{G}_3(Q)$ acts both on $\widetilde{Z}_n(Q,\delta)$ and on N in the natural way. Finally, the group $SO(4)$ acts in the obvious way on F by twisting the frame at s; it also acts in the natural way on $\widetilde{Z}_n(Q,\delta)/\mathcal{G}_3(Q)$ (see Problem 15).

By varying the bundle $Q \to S^4$, this construction allows us to describe all connections on P with exactly one point of concentrated curvature. In general, there are connections which have concentrated curvature at several points. There is a Taubes' gluing theorem which covers that situation as well. One specifies a finite set of small disjoint balls and gives a positive integer associated to each. This time choose $P' \to B$ so that the sum of these integers is the difference of $c_2(P) - c_2(P')$. One then glues in independently to each of these balls ASD connections from S^4 which are concentrated near the north pole, centered at the north pole, and have the energy specified by the integer associated to the ball.

Again, we can do this construction in non-redundant families so that, after dividing out by the appropriate automorphism groups, we obtain a diffeomorphism onto an open subset of $\mathcal{M}(P)$. The image of such a diffeomorphism is called a *Taubes neighborhood* and the diffeomorphism is called the *Taubes' gluing map*.

The point is that using Uhlenbeck's compactness result one can show that the images of these diffeomorphisms as we vary over all possible choices of bundles P' and number of balls and positive integers in fact covers a neighborhood of infinity in $\mathcal{M}(P)$. By this we mean that if $\{[A_n]\}$ is a sequence of points in $\mathcal{M}(P)$ then either there is a convergent subsequence in $\mathcal{M}(P)$ or there is a subsequence contained in one of the Taubes neighborhoods.

6.6. Negative definite 4-manifolds

Now let us apply these results to study smooth 4-manifolds with negative definite intersection forms.

Theorem 6.6.1. *Let B be a closed, oriented, simply connected smooth 4-manifold. Suppose that the intersection form on $H^2(B;\mathbf{Z})$ is negative definite. Then this intersection form is diagonalizable over $\mathbf{Z}$; that is to say, there is an integral basis $\{e_1, \dots, e_k\}$ for $H^2(B;\mathbf{Z})$ with the property that*

$$e_i \cdot e_j = -\delta_{i,j}.$$

This means that there is an orientation-reversing homotopy equivalence from B to $\#_k \mathbf{C}P^2$.

Proof. Let $P \to B$ be a principal $SU(2)$-bundle with $c_2(P) = 1$. Choose a generic metric g as in Theorem 5.3.1. Then the moduli space $\mathcal{M}^*(P,g)$ is a smooth manifold of dimension $8 \cdot 1 - 3(1 - 0 + 0) = 5$. Since $b_2^+(B) = 0$ there may well be reducible g-ASD connections. Applying Uhlenbeck's compactness result we see that a sequence of g-ASD connections has a subsequence which satisfies one of the following three properties:

- the subsequence converges to a point of $\mathcal{M}^*(P,g)$,
- the subsequence lies in a Taubes neighborhood of infinity, or
- the subsequence converges to a reducible g-ASD connection.

Thus, we see that $\mathcal{M}^*(P,g)$ has two types of ends. There is the end given by the Taubes' gluing theorem and the ends associated with reducible g-ASD connections. Let us analyze the Taubes end. The connections in this end have curvature concentrated in a small ball and elsewhere are almost flat, i.e., away from a small ball the connections are almost product connections. The concentrated part of the connection has energy approximately $8\pi^2$. For the trivial $SU(2)$-connection Θ the map $P_+\nabla_\Theta$ is identified with $P_+d\otimes \mathrm{Id}_{\mathfrak{g}}$. Since $b_2^+(B)=0$, we see that this map is surjective. Thus, the Taubes' gluing theorem applies to describe a neighborhood of any given connection of this type. The gluing parameters used to describe such connections are:

- a point $b\in B$ which is the center of mass of the concentrated curvature,
- an isomorphism between the fiber at the south pole of the principal bundle over S^4 with $c_2=1$ and the fiber of the trivial bundle over b,
- and an ASD connection on S^4 which is concentrated near the north pole and whose center of energy is the north pole.

Two sets of gluing parameters are equivalent if they differ by a global automorphism of the trivial connection on the trivial bundle on B and an automorphism of the principal bundle over S^4. Since the stabilizer of the trivial connection on B is $SU(2)$, it follows that the gluing parameters in this case are actually:

- a point $b\in B$ which is the center of mass of the connection, and
- an isomorphism class of ASD connections on S^4 which are concentrated near the north pole and centered at the north pole.

By Theorem 6.4.2, this space of isomorphism classes of ASD connections on S^4 is identified with an open interval $(1,\infty)$ where the parameter is a measure of the amount of concentration of the connection, as the parameter value goes to infinity the connection becomes more and more concentrated. This then gives us a diffeomorphism between $B\times(1,\infty)$ and the Taubes end of $\mathcal{M}^*(P)$. It is straightforward to show that this isomorphism is orientation-preserving. (Notice that since $b_2^+(B)=0$ our conventions have produced an orientation for $\mathcal{M}^*(P)$ without the necessity of making any choices related to B.)

Let us turn now to the analysis of the ends associated to reducible g-ASD connections.

Claim 6.6.2. *There is one isomorphism class of reducible g-ASD connections for each pair $\{\pm x\}$ where $x\in H^2(B;\mathbf{Z})$ and $x\cdot x=-1$.*

Proof. The pairs $\{\pm x\}$ as described in the claim are in natural one-to-one correspondence with the isomorphism classes of reductions of P to a principal S^1-bundle (see Problem 7). We need to see that each of these principal S^1-bundles Q has a unique g-ASD connection up to isomorphism. For each such bundle there is a connection whose curvature is the unique g-harmonic form representative for $c_1(Q)$. Since the intersection form on $H^2(B;\mathbf{Z})$ is negative definite this harmonic form must be g-ASD. This shows that there is a reducible g-ASD connection for each pair $\{\pm x\}$ as in the statement of the claim.

Let us show that the reducible g-ASD connection associated to $\{\pm x\}$ is unique up to isomorphism. Equivalently, we must show that the principal S^1-bundle Q has a unique g-ASD connection up to isomorphism. We know that the curvature of any such connection must be a closed g-ASD form representing $c_1(Q)$. Any such form is harmonic (see Lecture V Problem 37) – consequently, all g-ASD connections on Q have the same curvature. It is now an easy exercise (see Problem 6) to show that since B is simply connected two connections on Q with the same curvature form differ by an automorphism of Q.

It now follows that there is, up to automorphisms of P, exactly one reducible g-ASD connection on P for each pair $\{\pm x\}$ as in the statement of the claim. □

Now we need to understand the structure of the end of $\mathcal{M}^*(P,g)$ near a given reducible ASD connection. Let A be a reducible g-ASD connection. Let $U \subset (\{A\} + \mathrm{Ker}\nabla_A^*) \subset \mathcal{A}_2(P)$ be the slice for the action of $\mathcal{G}_3(P)$. Since g has been chosen in the G_δ given in Theorem 5.3.1, we see that

$$P_+F \colon U \to \Omega^2_+(B;\mathrm{ad}P)_{L^2_1}$$

is transverse to zero at A with preimage $\widetilde{\mathcal{M}}(P,g)\cap U$. Hence, if we choose U sufficiently small, the intersection $\widetilde{\mathcal{M}}(P,g)\cap U$ is a smooth manifold. Its dimension is the dimension of $\mathrm{Ker}(P_+\nabla_A|_{\mathrm{Ker}\nabla_A^*}) = H^1(\mathcal{E}(A))$. Since $H^0(\mathcal{E}(A))$ is the tangent space to $\mathrm{Stab}(A) = S^1$, it has dimension one. On the other hand, the Euler characteristic $\chi(\mathcal{E}(A)) = -5$. Since for generic g we have $H^2(\mathcal{E}(A)) = 0$; it follows that the dimension of $\widetilde{\mathcal{M}}(P,g)\cap U$ is 6. This six-manifold has an action of $S^1 = \mathrm{Stab}(A)$ on it. Since all reducible g-ASD connections sufficiently near to A are isomorphic to A and since U/S^1 embeds into the quotient space, we see that the only reducible connection in $\widetilde{\mathcal{M}}(P,g)\cap U$ is A. This means that the action of S^1 on this 6 manifold has an isolated fixed point at A. All other points have stabilizer $\{\pm 1\}$. Hence, dividing by $\{\pm 1\}$ we have a semi-free smooth action of S^1 on $\widetilde{\mathcal{M}}(P,g)\cap U$ with an isolated fixed point at A. It follows that near A the action is differentiably equivalent to the standard action of S^1 by complex multiplication on $\{z \in \mathbf{C}^3 \big| \|z\| < \epsilon\}$. Of course the quotient of this action embeds in $\mathcal{B}(P)$ homeomorphically onto a neighborhood of $[A]$ in $\mathcal{M}(P,g)$. It follows that the end of $\mathcal{M}^*(P,g)$ is diffeomorphic to $\mathbf{C}P^2 \times (0,1)$ where, as the interval parameter goes to 0, the connections are converging to the reducible connection. It is also true that the orientation of $\mathcal{M}^*(P,g)$ is the opposite orientation from that of $\mathbf{C}P^2 \times (0,1)$, though we do not need to establish this directly.

Cutting off open neighborhoods of infinity, we obtain a compact submanifold with boundary $M \subset \mathcal{M}^*(P,g)$. This compact manifold is oriented and its boundary is

$$B \cup \coprod_{\{\pm x\}} \mathbf{C}P^2$$

where the pairs range over all $\{\pm x\} \subset H^2(B;\mathbf{Z})$ such that $x \cdot x = -1$. The orientation of M induces the usual orientation on B but we have not attempted to identify the orientation on the $\mathbf{C}P^2$ components. The signature of the boundary is zero. For each pair $\{\pm x\}$ let $\epsilon_{\{\pm x\}} = \pm 1$ be the signature of corresponding boundary component of M taken with the induced orientation. Thus, we have

$$\sigma(B) + \sum_{\{\pm x\}} \epsilon_{\{\pm x\}} = 0.$$

Since B is negative definite, we see that $\sigma(B) = -\text{rank}H^2(B)$. It follows that the number of pairs $\{\pm x\} \subset H^2(B;\mathbf{Z})$ with $x \cdot x = -1$ is at least the rank of $H^2(B)$. According to Problem 14 of Lecture I, it follows that the intersection form of B is negative definite. Notice that it also follows that $\epsilon_{\{\pm x\}} = +1$ for every $\{\pm x\}$. In turn, this shows that the orientation near the reducible ends is the usual one on $\mathbf{C}P^2$. □

6.7. Problems

1. Which intersection forms occur as intersection forms of closed, oriented, smooth, positive definite 4-manifolds? Why?
2. Let Σ be the Poincaré homology 3-sphere, i.e., $\Sigma = \partial W$ where W is the four-manifold constructed from plumbing tangent disk bundles of S^2 as indicated by the Dynkin diagram for E_8. Show that for any $n > 0$ the (oriented) connected sum Σ_n of n copies of Σ does not bound a contractable manifold. Show that Σ_n does not bound a homology disk whose fundamental group is normally generated by the fundamental group of Σ_n.
3. Establish the same results as in Problem 2 for any homology sphere which bounds a simply connected, smooth manifold with a positive definite intersection form.
4. Show that no homology sphere can bound a both a positive definite simply connected 4-manifold with non-diagonalizable intersection form and a negative definite simply connected 4-manifold.
5. Show the $K3$ surface can not be decomposed as a connected sum
$$X\#(S^2 \times S^2)\#(S^2 \times S^2)\#(S^2 \times S^2).$$
6. Show that if $Q \to B$ is a principal S^1-bundle over a simply-connected base B, then two connections on Q with the same curvature form differ by an automorphism of Q.
7. Show that the set of isomorphism classes of S^1-reductions of an $SU(2)$-bundle P over a closed 4-manifold B is in one-to-one correspondence with the set of pairs $\{\pm x\} \subset H^2(B;\mathbf{Z})$ with the property that $x^2 = -c_2(P)$.
8. Show that if B has a riemannian metric with positive sectional curvature then for any ASD connection A we have that $P_+\nabla_A$ is surjective. Hint: The condition on the curvature implies that the restriction of Δ_A to the self-dual two-forms is a positive operator, and in particular has no kernel (see [**4**]).
9. Identify S^7 with the unit vectors in $\mathbf{H} \times \mathbf{H}$, under the norm
$$|(q_1, q_2)| = q_1\bar{q}_1 + q_2\bar{q}_2.$$
The sphere S^7 admits a free right-action by $SU(2)$, the unit quaternions, given by
$$(q_1, q_2) \times h \mapsto (q_1 h, q_2 h).$$
Show that the quotient space, $\mathbf{HP}^1$ is naturally identified with S^4. Show that c_2 of this bundle is 1, giving us an explicit model for the bundle Q of Chern class 1 over S^4.

10. Show that the natural left action of $Sp(2)$ on this S^7 projects to an action of $SU(2)\backslash Sp(2)$ on $\mathbf{HP}^1$. By analogy with the Möbius group acting on the Riemann sphere, we write elements of $SU(2)\backslash Sp(2)$ as functions
$$q \mapsto (aq+b)(-\bar{b}q+\bar{a})^{-1}.$$
Show that these transformations are conformal and that, indeed, any conformal transformation of S^4 can be written in this way; i.e. $\mathrm{Conf}(S^4) \cong SU(2)\backslash Sp(2)$.
11. Show that $Sp(2)$ naturally acts on the space of connections over Q.
12. We have shown in Problem 9 that the fibers of $S^7 \to \mathbf{HP}^1 = S^4$ are the orbits under right multiplication by unit quaternions. Use the natural metric to define an orthogonal projection from the tangent bundle of S^7 to the vertical tangents. Show that this induces a connection over Q. Prove that this connection is ASD. (Hint: check at a convenient point and then use the $Sp(2)$ action.)
13. Show that the $SU(2)$ stabilizer of a point in S^4 under the $Sp(2)$ action can be viewed as a gauge transformation of the bundle, hence that the $Sp(2)$ action from the previous problem descends to give an action of $\mathrm{Conf}(S^4)$ on $\mathcal{B}(Q)$.
14. Show that $p_1(\Lambda^2_- T^*S^4) = -4$, and hence that it can be written as adQ, where $c_2(Q) = 1$. Show that there is a natural action of $\mathrm{Diff}(S^4)$ on $\Lambda^2_- T^*S^4$ and on ad$Q$. Show that the action on ad$Q$ commutes with the right $SO(3)$ action, and hence that $\mathrm{Diff}(S^4)$ (hence, *a fortiori* $\mathrm{Conf}(S^4)$) acts on the space $\mathcal{A}(Q)$. Does the $\mathrm{Conf}(S^4)$-action lift to one on Q?
15. Show that $\mathrm{Diff}(B)$ acts naturally on the space $\mathcal{B}(P)$. Show that the restriction of this action to the conformal subgroup $\mathrm{Conf}(B)$ leaves invariant the subset $\mathcal{M}(P)$.
16. Let $Q \to S^4$ be the principal $SU(2)$ bundle with $c_2 = 1$ bundle. Show that the orbits of the conformal group action on $\mathcal{M}(Q)$ (with S^4 given the round metric) are closed.
17. Define the space of based connections to be the quotient
$$\mathcal{A}_2(P) \times_{\mathcal{G}_3(P)} P.$$
Show that this is a Hilbert manifold. What is the fiber of the natural mapping
$$\mathcal{A}_2(P) \times_{\mathcal{G}_3(P)} P \to \mathcal{B}(P) \times B$$
over $([A], b)$? Show that the restriction of this mapping over $\mathcal{B}_2^*(P) \times B$ is a principal $SO(3)$-bundle. Define the based moduli space to be
$$\widetilde{\mathcal{M}}(P) \times_{\mathcal{G}_3(P)} P.$$
Show that for a generic metric this is a smooth manifold of dimension $d(P) + 7$.
18. Fix a point $b \in B$. We define the space of gauge equivalence classes of based connections based at b to be $\mathcal{A}_2(P) \times_{\mathcal{G}_3(P)} P_b$. Show that this is a Hilbert manifold. We define the corresponding moduli space to be based at b by setting ${}^b\mathcal{M}(P) = \widetilde{\mathcal{M}}(P) \times_{\mathcal{G}_3(P)} P_b$. Show that for a generic metric g, this moduli space based at b is a smooth manifold of dimension $d(P) + 3$. Show that the projection ${}^b\mathcal{M}^*(P) \to \mathcal{M}^*(P)$ is the projection mapping of a

principal $SO(3)$-bundle. Given an equivalence class of connections $[A]$ (not necessarily ASD), show

$$\mathrm{Stab}_{SU(2)}{}^b[A] \cong \mathrm{Stab}_{\mathcal{G}} A.$$

19. Show that the space of gluing data for Taubes' gluing theorem 6.4.4 is isomorphic to

$${}^xU \times {}^{x'}U'/SO(3) \subset {}^x\mathcal{M}(P) \times {}^{x'}\mathcal{M}(P')/SO(3).$$

20. Suppose that $P \to B$ and $P' \to B'$ are principal G-bundles. Suppose that $V \subset B$ and $V' \subset B'$ are geodesic balls. Define the connected sum $B\#B'$ of B and B' in terms of an inversion isomorphism from an annulus in V to one in V'. Show that for given trivializations of $P|_V$ and $P'|_{V'}$ and an isomorphism $\sigma\colon P_x \to P'_{x'}$ there is a natural way to construct a principal G-bundle $P\#P'$ over $B\#B'$. Show that this bundle agrees with P over $B-V$ and with P' over $B'-V'$. Show that the resulting connected sum bundle is independent, up to isomorphism, of the choice of trivializations. Suppose that P and P' have connections A and A' which are product connections over V and V' respectively. Show that if we use trivializations of $P|_V$ and $P'|_{V'}$ in which A and A' are product connections, then there is a natural connection $A\#A'$ on $P\#P'$ which agrees with A over $B-V$ and with A' over $B'-V'$. Is the gauge equivalence class of $A\#A'$ independent of σ?
21. Following notation from the previous exercise, show that there is a natural inclusion

$$\mathrm{Stab}_{\mathcal{G}(P\#P')}(A\#A') \subset \mathrm{Stab}_{\mathcal{G}(P)}(A) \times \mathrm{Stab}_{\mathcal{G}(P')}(A').$$

Indeed, show that the elements of $\mathrm{Stab}_{\mathcal{G}(P\#P')}(A\#A')$ correspond exactly to those pairs of gauge transformations which stabilize σ.
22. Show that the gluing map defined on the level of connections in Corollary 6.3.3 descends to a map on gauge equivalence classes as stated in that corollary.
23. Let A be a connection on a principal bundle $P \to B$. Let $V \subset B$ be a geodesic ball centered at $x \in B$. Show that parallel translation along geodesics radiating form x determines a product structure for $P|_V$.
24. Let A be a connection on $P \to B$ and let $V' \subset V$ be concentric geodesic balls in B. Show that there is a constant C depending only on V, V' such that if $P|_V$ has a product structure in which $A|_V$ is of the form $d+\alpha$ with $\|\alpha\|_{L^2_2} < \epsilon$ then then there is a connection A_1 on P with $\|A-A_1\|_{L^2_2} < C\epsilon$ which agrees with A away from V and which is a product connection in the given product structure over the geodesic ball $V' \subset V$. Hint: Let ψ be a C^∞-function on B with the support of $1-\psi$ contained in V and with $\psi|_{V'} = 0$. Form the connection A_1 which is given by A over $B-V$ and is given by $\Theta + \psi\alpha$ over V where Θ is the product connection.

LECTURE 7
The Donaldson Polynomial Invariants

In this lecture we will sketch the definition of the Donaldson polynomial invariants of a smooth 4-manifold. Throughout this lecture B is a closed oriented, simply connected smooth 4-manifold, with a fixed orientation of $H^2_+(B;\mathbf{R})$. For the rest of this lecture we shall assume that $b_2^+(B) > 1$.

7.1. The formalism of the Donaldson polynomial invariants

Before we begin with the definition proper, let us describe the type of objects that we will eventually define.

For each principal $SU(2)$-bundle P with $c_2(P) > 0$ we set $d(P)$ to be equal to one-half the value it had previously. Namely

$$d = d(P) = 4c_2(P) - \frac{3}{2}(1 + b_2^+(B)).$$

This number is an integer provided that $b_2^+(B)$ is odd. From now on we shall implicitly make this assumption. Notice that d is one-half the formal dimension of the moduli space $\mathcal{M}^*(P)$. The Donaldson polynomial invariant associated to P is a symmetric multilinear function of degree d on $H_2(B;\mathbf{Z})$ with values in $\mathbf{Q}$:

$$\gamma_d(B) = \gamma_d \colon \underbrace{H_2(B;\mathbf{Z}) \otimes \cdots \otimes H_2(B;\mathbf{Z})}_{d \text{ times}} \to \mathbf{Q}.$$

Equivalently, we can view γ_d as a homogeneous polynomial function on $H_2(B;\mathbf{Z})$. It turns out that these functions have strong integrality properties. If $c_2(P)$ is sufficiently large (depending on $b_2^+(B)$), then the values of γ_d are integral. In general its values are always of the form $a/2^r$ with a an integer; the power of 2 in the denominator can be bounded above in terms of $c_2(P)$ and $b_2^+(B)$.

The basic property of these functions is that each γ_d is an orientation-preserving diffeomorphism invariant of B together with the orientation on $H^2_+(B;\mathbf{R})$. If we reverse the orientation of $H^2_+(B;\mathbf{R})$ then the invariant changes sign. If we reverse the orientation of B then we do violence to these invariants. In most examples these invariants vanish for one of the two possible orientations of B.

The first manifold for which these invariants were computed and non-trivial was the $K3$ surfaces. Let B be a $K3$ surface with its natural complex orientation. We orient $H^2_+(B;\mathbf{R})$ as follows. There is an identification of this topological cohomology with $H^{2,0}(B;\mathbf{C})\oplus\mathbf{R}(\omega)$, where ω is the Kähler class. This vector space is oriented by using the complex orientation on the first factor and using ω to orient the second factor. Following Kronheimer, we form the formal power series on $H_2(B;\mathbf{R})$, called the *Donaldson series* of B

$$D_B(x)=\sum_{d\geq 0}\gamma_d(x)/d!.$$

This power series converges absolutely and defines the analytic function given by the even terms in the power series

$$\exp(q_B/2),$$

where q_B is the quadratic polynomial on $H_2(B;\mathbf{R})$ which defines the intersection pairing; i.e., $q_B(x)=x\cdot x$. Notice that this is just a compact way of saying that

$$\gamma_d(B)(x)=\frac{q_B(x)^{d/2}}{2^d d!}.$$

(See Problem 1.) This computation was done in [**5**]. If B' is the blow up the $K3$ at one point, then $D_{B'}(x)$ is the analytic function given by the even degree terms in $\exp(q_{B'}/2)\cosh(e)$, where similarly to before $q_{B'}$ is the quadratic intersection form for B', and e is the cohomology class Poincaré dual to the class of the exceptional curve. This computation was first done by Kronheimer and Mrowka [**7**].

There are by now many other manifolds for which the Donaldson series have been computed. All the known computations yield answers which are of the same general nature as this one – the Donaldson series converges absolutely for all x and yields functions expressible in terms of the exponentials of q and certain linear classes. For more on this aspect, see Stern's lectures.

In order to define the Donaldson polynomial invariants, we first define a map μ from the polynomial algebra generated by the rational homology of B to the cohomology ring of $\mathcal{B}^*(P)$. The rough idea is to integrate the resulting classes over the fundamental class of the moduli space of ASD connections in order to give an element in the dual to this polynomial algebra, which is simply the polynomial algebra on the cohomology of B. The resulting element is the Donaldson polynomial invariant. Unfortunately, the moduli space is not compact. Nevertheless, restricting the map μ to the moduli space induces a map from the polynomial algebra on the homology of B to the cohomology ring of the moduli space. Then we define a compactification of the moduli space and extending the map from the polynomial algebra on the homology of B into the cohomology ring of this compactification. Then we are in a position to integrate over the fundamental cycle in order to define the Donaldson invariants.

7.2. The μ-map

The purpose of this section is to define the μ-map

$$\mu\colon H_2(B;\mathbf{Z})\to H^2(\mathcal{M}^*(P);\mathbf{Z}).$$

This is an essential ingredient in the definition of the Donaldson invariants $\gamma_d(B)$. Recall that there is a principal $SO(3)$-bundle

$$\mathcal{A}^*_2(P)\times_{\mathcal{G}_3(P)}P\to\mathcal{B}^*(P)\times B.$$

We give the name ξ to the total space of this principal bundle. Then we have the class $p_1(\xi)$ and slant product with this class defines a homomorphism

$$H_2(B;\mathbf{Z}) \to H^2(\mathcal{B}_2^*(P);\mathbf{Z})$$

by

$$x \mapsto p_1(\xi)/x.$$

Restricting this map to $\mathcal{M}^*(P) \subset \mathcal{B}_2^*(P)$ gives us a map

$$\mu\colon H_2(B;\mathbf{Z}) \to H^2(\mathcal{M}^*(P);\mathbf{Z}).$$

Lemma 7.2.1. *This map is divisible by* 4.

Proof. For any $SO(3)$-bundle ζ, the first Pontrjagin class $p_1(\zeta)$ reduced modulo 4 is equal to $\mathcal{P}(w_2(\zeta))$ where $\mathcal{P}$ is the Pontrjagin square. For any $A \in \mathcal{A}_2^*(P)$ the restriction $\xi|_{(\{[A]\},B)}$ lifts to an $SU(2)$-bundle. This implies that $w_2(\xi)$ is pulled back from $\xi|_{\mathcal{B}_2^*(P)}$ by projection to the first factor. The same is of course true for $\mathcal{P}(w_2(\xi))$. Thus, $\mathcal{P}(w_2(\xi))/x = 0$ for any $x \in H_2(B;\mathbf{Z})$. This proves that $p_1(\xi)/x \equiv 0 \pmod 4$ for any $x \in H_2(B;\mathbf{Z})$. □

Corollary 7.2.2. *We define*

$$\mu(x) = \frac{-p_1(\xi)/x}{4}.$$

This gives a map

$$\mu\colon H_2(B;\mathbf{Z}) \to H^2(\mathcal{B}_2^*(P);\mathbf{Z})$$

and by restriction a map

$$\mu\colon H_2(B;\mathbf{Z}) \to H^2(\mathcal{M}^*(P);\mathbf{Z}).$$

Remark 7.2.3. If the bundle ξ could be lifted to a principal $SU(2)$-bundle, then $\mu(x)$ would be the slant product of x with c_2 of this lifted bundle (see Problem 18).This is indeed the case when B is simply connected and $c_2(P)$ is odd. In the case when c_2 is even the bundle does not lift to an $SU(2)$ bundle, but still $p_1(\xi)$ is divisible by 4.

There is a similar map

$$\mu\colon H_0(B;\mathbf{Z}) \to H^4(\mathcal{B}_2^*(P);\mathbf{Z})$$

given by sending $1 \in H_0(B;\mathbf{Z})$ to $(-1/4)p_1(\xi)$. As we just remarked, this map is integral.

Proposition 7.2.4. *Suppose that $H_1(B;\mathbf{R}) = 0$. Then the rational cohomology ring of $\mathcal{B}^*(P)$ is a polynomial algebra generated by the images of the μ-map. That is to say*

$$H^*(\mathcal{B}_2^*(P);\mathbf{Q}) = S^*(\mu(H_2(B;\mathbf{Q}))) \otimes S^*(\mu(H_0(B;\mathbf{Q}))),$$

where $S^*(V)$ denotes the symmetric algebra generated by V. The usual grading of this algebra is induced by defining the vector space $\mu(H_i(B;\mathbf{Q}))$ to be homogeneous of degree $4-i$.

7.3. The Uhlenbeck compactification of the moduli space

Uhlenbeck's compactness result for $\mathcal{M}(P)$ leads us naturally to a compactification of the moduli space. This compactification was introduced by Donaldson, but since it is based on Uhlenbeck's results it is usually called the *Uhlenbeck compactification.*

Any ASD L^2_2-connection A on a principal bundle $P \to B$ determines an L^1-function $|F_A|^2$ on B. Multiplying this function by the volume element of B gives us a measure on B of total mass $8\pi^2 c_2(P)$. Let $\Sigma^t(B)$ denote the tth symmetric product of B, the space of t unordered and possibly non-distinct points in B. Given a point $([A], \{x_1, \dots, x_t\}) \in \mathcal{M}(P) \times \Sigma^t(B)$ we define an associated measure

$$|F_A|^2 d\mathrm{Vol} + \sum_{i=1}^{t} 8\pi^2 \delta_{x_i}$$

where δ_{x_i} is the δ-mass of mass one supported at x_i. The integral of this measure is $8\pi^2(c_2(P) + t)$.

Let $P_k \to B$ be a principal $SU(2)$-bundle with $c_2(P_k) = k$. To define the Uhlenbeck compactification of $\mathcal{M}(P_k)$ we take

$$\coprod_{t=0}^{k} \mathcal{M}(P_{k-t}) \times \Sigma^t(B).$$

We define a topology on this disjoint union as follows: a sequence of points $\{([A_n], \{x_1^n, \dots, x_{t(n)}^n\})\}$ converges to $([A], \{x_1, \dots, x_t\})$ if

- The measures μ_n associated to the points of the sequence converge weakly to the measure for the putative limit point.
- For every compact subset $K \subset B - \{x_1, \dots, x_t\}$ the restrictions of the connections A_n to this set converge, up to changes of gauge, in the L^2_2-topology to the restriction of A.

Notice that if a sequence $\{([A_n], \{x_1^n, \dots, x_{t(n)}^n\})\}$ converges to $([A], \{x_1, \dots, x_t\})$ then the sets $S_n \subset B$ defined by $S_n = \{x_1^n, \dots, x_{t(n)}^n\}$ converge to a subset of $\{x_1, \dots, x_t\}$. We denote by $\overline{\mathcal{M}}(P_k)$ the resulting topological space. The points of $\overline{\mathcal{M}}(P_k)$ are called *generalized ASD connections on* P_k. The connection A is called the *background connection* of any generalized ASD connection of the form $([A], \{x_1, \dots, x_t\})$. The subset $\mathcal{S} = \{x_1, \dots .x_t\} \subset M$ is called its *singular locus.* The measure

$$|F_A|^2 d\mathrm{Vol} + \sum_{i=1}^{t} 8\pi^2 n_i \delta_{x_i}$$

is called the *measure* of the generalized ASD connection.

The following lemma is an easy consequence of Uhlenbeck's compactness theorem.

Lemma 7.3.1. *The space $\overline{\mathcal{M}}(P_k)$ is a compact, Hausdorff space. It contains $\mathcal{M}(P_k)$ as an open subset.*

In general, it is not the case that $\mathcal{M}(P_k)$ is dense in $\overline{\mathcal{M}}(P_k)$ so that strictly speaking $\overline{\mathcal{M}}(P_k)$ is not a compactification of $\mathcal{M}(P_k)$. It is true, however, that $\mathcal{M}(P_k)$ is dense in $\overline{\mathcal{M}}(P_k)$ if k is sufficiently large (how large depending on $b_2^+(B)$).

7.4. Extension of the μ-map over the compactification

We chose a metric g on B for which the moduli spaces $\mathcal{M}^*(P)$ are smooth for every non-trivial principal $SU(2)$-bundle $P \to B$. Since $b_2^+(B) > 0$, we can also arrange that there are no reducible g-ASD connections on P either. This means that if P is non-trivial then $\mathcal{M}^*(P) = \mathcal{M}(P)$ is a smooth manifold of dimension $2d(P)$.

The μ-map that we have defined is not sufficient for the applications because of the non-compactness of $\mathcal{M}(P)$. To remedy this problem we extend the μ-map to a map

$$\overline{\mu}\colon H_2(B;\mathbf{Z}) \to H^2(\overline{\mathcal{M}}(P);\mathbf{Z}).$$

That is the purpose of this subsection.

Theorem 7.4.1. *There exists a natural map*

$$\overline{\mu}\colon H_2(B;\mathbf{Z}) \to H^2(\overline{\mathcal{M}}(P);\mathbf{Z})$$

which when followed by the restriction mapping

$$H^2(\overline{\mathcal{M}}(P);\mathbf{Z}) \to H^2(\mathcal{M}(P);\mathbf{Z})$$

is equal to μ.

We shall not give the details of the proof of this result. We content ourselves with outlining the proof. First of all, we define maps

$$\mu_t\colon H_2(B;\mathbf{Z}) \to H^2(\mathcal{M}(P_{k-t}) \times \Sigma^t(B);\mathbf{Z})$$

by

$$\mu_t(x) = \mu(x) + \Sigma^t(\mathrm{PD}^{-1}(x))$$

where PD is the Poincaré duality isomorphism and for any $y \in H^2(B;\mathbf{Z})$ the class $\Sigma^t(y) \in H^2(\Sigma^t(B);\mathbf{Z})$ is the class which pulls back to $\underbrace{B \times \cdots \times B}_{t\text{ times}}$ to give the class

$$(y \otimes 1 \otimes \cdots \otimes 1) + (1 \otimes y \otimes 1 \cdots \otimes 1) + \cdots + (1 \otimes \cdots \otimes 1 \otimes y).$$

It turns out that there is a unique map

$$\overline{\mu}\colon H_2(B;\mathbf{Z}) \to H^2(\overline{\mathcal{M}}(P);\mathbf{Z})$$

whose restriction to the closed subset $\mathcal{M}(P_{k-t}) \times \Sigma^t(B)$ is equal to μ_t, for every t.

To prove this extension result we need to understand how the various pieces making up $\overline{\mathcal{M}}(P)$ fit together. The basic fact is that $\overline{\mathcal{M}}(P)$ is a stratified space[1]. This is a consequence of the Taubes parameters for neighborhoods of infinity. There is one other fact that is needed, namely that the moduli spaces of ASD connections on S^4 are all non-empty. This of course follows from the Chern class one bundle by direct construction and follows for the higher charge bundles by Taubes' gluing theorem. With these facts, one easily establishes that there is a unique map $\overline{\mu}$ restricting to give the μ_t.

A similar construction works for $\mu(1)$ as well except near the subset of $\overline{\mathcal{M}}(P)$ whose background connection is trivial.

[1]Strictly, speaking this is not known to be true near the points of $\overline{\mathcal{M}}(P)$ with trivial background connection.

Proposition 7.4.2. *Let $T \subset \overline{\mathcal{M}}(P)$ be the closed subset consisting of generalized ASD connections with trivial background connection. Then there is a unique class*

$$\overline{\mu}(1) \in H^4(\overline{\mathcal{M}}(P) - T; \mathbf{Z})$$

whose restriction to $\mathcal{M}(P_{k-t}) \times \Sigma^t(B)$, for any $t < k$, is equal to

$$\mu(1) + \Sigma^t([B]^*)$$

where $[B]^* \in H^4(B; \mathbf{Z})$ is the fundamental cohomology class of B and $\Sigma^t([B]^*)$ is the obvious symmetrization of this class.

It is definitely not the case that this class extends over T. Its evaluation on the link of a generic point of T is a class of infinite order in the fourth cohomology of the link.

7.5. Definition of the Donaldson polynomial invariants in the stable range

Consider the compactified moduli space $\overline{\mathcal{M}}(P_k)$. Notice that the dimension of the subset $\mathcal{M}(P_{k-t}) \times \Sigma^t(B)$ is equal to $2d(P_k) - 4t$ unless $t = k$. The dimension of $\Sigma^k(B) = 4k = 2d(P_k) - 4k + 3(1 + b_2^+(B))$. The reason for this anomaly is that the moduli space of the trivial bundle is a single point, yet its formal dimension is $-3(1 + b_2^+(B))$. In order that $\overline{\mathcal{M}}(P_k)$ have a fundamental class, it is imperative that all the pieces except $\mathcal{M}(P_k)$ have dimension which is at least two less than that of $\mathcal{M}(P_k)$. This is automatic except for $\Sigma^k(B)$. For this condition to hold for this last piece we need

$$4k \geq 3b_2^+(B) + 5.$$

If this condition holds then k is said to be *in the stable range*.

Proposition 7.5.1. *Let B be a closed, oriented, simply connected 4-manifold with $b_2^+(B) > 0$. Fix an orientation for $H_+^2(B; \mathbf{R})$. Let $P \to B$ be a principal $SU(2)$-bundle. If $c_2(P)$ is in the stable range, then for any generic metric g on B the compactified moduli space $\overline{\mathcal{M}}(P)$ has a fundamental class*

$$[\overline{\mathcal{M}}(P)] \in H_{2d(P)}(\overline{\mathcal{M}}(P); \mathbf{Z}).$$

The locally finite class induced by restriction to $\mathcal{M}(P)$ is the orientation class for the manifold $\mathcal{M}(P)$ with orientation induced by the orientation of $H_+^2(B; \mathbf{R})$. If $b_2^+(B) > 1$, and if g_0 and g_1 are generic metrics, then for a generic path of metrics $\{g_t\}$ joining g_0 to g_1 the parametrized compactified moduli space $\overline{\mathcal{M}}(P, \{g_t\})$ has a relative fundamental class whose boundary is the difference

$$[\overline{\mathcal{M}}(P, g_1)] - [\overline{\mathcal{M}}(P, g_0)].$$

Now we are in a position to define

$$\gamma_d(B)\colon \underbrace{H_2(B; \mathbf{Z}) \otimes \cdots \otimes H_2(B; \mathbf{Z})}_{d \text{ times}} \to \mathbf{Z},$$

when $b_2^+(B) > 1$, when $c_2(P)$ is in the stable range, and when d is congruent to $-3(1 + b_2^+(B))/2$ modulo 4. First, fix a principal $SU(2)$-bundle with $d(P) = d$. Then, we choose a generic riemannian metric on B, one for which $\mathcal{M}(P') = \mathcal{M}^*(P')$

is a smooth manifold of the expected dimension for every non-trivial principal $SU(2)$-bundle $P' \to B$. For any classes $x_1, \ldots, x_d \in H_2(B; \mathbf{Z})$ we set

$$\gamma_d(x_1, \ldots, x_d) = \langle \overline{\mu}(x_1) \cup \cdots \cup \overline{\mu}(x_d), [\overline{\mathcal{M}}(P)] \rangle.$$

This is clearly a multilinear map. The only thing to prove is that it is independent of the generic metric g. But if g_0, g_1 are generic riemannian metrics on B, then since $b_2^+(B) > 1$ we can joint them by a path $\{g_t\}$ of metrics for which $\mathcal{M}(P, \{g_t\}) = \mathcal{M}^*(P, \{g_t\})$ is a smooth manifold. We then form the compactified parametrized moduli space $\overline{\mathcal{M}}(P, \{g_t\})$. This is a compact space whose boundary is $\overline{\mathcal{M}}(P, g_1) - \overline{\mathcal{M}}(P, g_0)$. There is a map

$$\overline{\mu}\colon H_2(B; \mathbf{Z}) \to H^2(\overline{\mathcal{M}}(P, \{g_t\}); \mathbf{Z})$$

extending the $\overline{\mu}$ maps for the two ends. It follows easily from this and the existence of the relative fundamental class for the parametrized moduli space that the function γ_d defined using g_0 and g_1 are equal. This proves that γ_d is independent of the generic riemannian metric g on B. We take this as the definition of $\gamma_d(B)$.

In exactly the same way we can define functions

$$\gamma_d(B)\colon \underbrace{H_2(B; \mathbf{Z}) \otimes \cdots \otimes H_2(B; \mathbf{Z})}_{a \text{ times}} \otimes \underbrace{H_0(B; \mathbf{Z}) \otimes \cdots \otimes H_0(B; \mathbf{Z})}_{b \text{ times}} \to Z$$

for $a + 2b = d$ and $4a \geq 3b_2^+(B) + 5$.

Of course the case when $b = 0$ is just the case considered above.

We have established the following theorem.

Theorem 7.5.2. *Let B be a closed oriented, simply connected four-manifold with $b_2^+(B) > 1$. Then for any $a, b \geq 0$ with $d = a + 2b \equiv -3(b_2^+(B) + 1)/2 \pmod 4$ with $4a \geq 3b_2^+(B) + 5$ the map*

$$\gamma_d(B)\colon \underbrace{H_2(B; \mathbf{Z}) \otimes \cdots \otimes H_2(B; \mathbf{Z})}_{a \text{ times}} \otimes \underbrace{H_0(B; \mathbf{Z}) \otimes \cdots \otimes H_0(B; \mathbf{Z})}_{b \text{ times}} \to Z$$

defined above is a C^∞-invariant of B and is called the stable Donaldson Polynomial Invariant *of B.*

7.6. A blow-up formula and the unstable range

In this section we wish to remove the stable range restriction encountered in the previous theorem. The idea is a simple one and relies on a blow-up formula.

Proposition 7.6.1. *Suppose that d is in the stable range for the manifold B. Let B' be the blow-up of B, i.e., the connected sum of B and $\overline{\mathbf{C}P}^2$. Then $d+1$ is in the stable range for B'. Let $e \in H_2(B'; \mathbf{Z})$ be the class represented by the exceptional curve in B', i.e., the generator of $H_2(\overline{\mathbf{C}P}^2; \mathbf{Z})$. Then for any classes $x_1, \ldots, x_d \in H_2(B; \mathbf{Z})$ we have*

$$\gamma_{d+1}(B')(x_1, \ldots, x_d, e, e, e, e) = -2\gamma_d(B)(x_1, \ldots, x_d).$$

The twist in this proposition is that $d + 1$ can be in the stable range for B' without d being in the stable range for B. In fact, since the stable range for a manifold depends only on b_2^+ of that manifold, B and B' have the same stable range. This leads us to the following result.

Proposition 7.6.2. *Let B and $d \geq 0$ be given. Suppose that $b_2^+(B) > 1$. Then there is $b > 0$ such that $d + b$ is in the stable range for the b-fold blow up B' of B. Let $e_1, \ldots, e_b \in H_2(B' : \mathbf{Z})$ be the classes represented by the exceptional curves in B'. For any classes $x_1, \ldots, x_d \in H_2(B; \mathbf{Z})$ we consider*

$$\left(\frac{-1}{2}\right)^b \gamma_{d+b}(B')(x_1, \ldots, x_d, e_1, \ldots, e_b)$$

where each e_i occurs 4 times. The result is independent of b, provided $d + b$ is in the stable range for B, and defines a multilinear symmetric function

$$\gamma_d(B)\colon \underbrace{H_2(B;\mathbf{Z}) \otimes \cdots \otimes H_2(B;\mathbf{Z})}_{d \text{ times}} \to \mathbf{Q}$$

whose image is contained in

$$\left(\frac{1}{2}\right)^{b_0} \mathbf{Z} \subset \mathbf{Q}$$

where b_0 is the minimal integer with the property that $d + b_0$ is in the stable range.

We can extend $\gamma_d(B)$ similarly to

$$\gamma_d(B)\colon \underbrace{H_2(B;\mathbf{Z}) \otimes \cdots \otimes H_2(B;\mathbf{Z})}_{a \text{ times}} \otimes \underbrace{H_0(M;\mathbf{Z}) \otimes \cdots \otimes H_0(M;\mathbf{Z})}_{b \text{ times}} \to \mathbf{Q}$$

for $a + 2b = d$. Once again there is a direct estimate on the denominators of the image of this map.

This is the *full Donaldson Polynomial.*

7.7. The Donaldson series

By definition we can define $\gamma_d(B)$ only when $d \equiv 3(1 + b_2^+(B))/2 \pmod 4$. There is a nice trick introduced by Kronheimer to define $\gamma_d(B)$ as long as $d \equiv 3(1 + b_2^+(B))/2 \pmod 2$. For those values of d satisfying this congruence but not the mod 4 congruence we set

$$\gamma_d(B)(x_1, \ldots, x_d) = \frac{1}{2}\gamma_{d+2}(B)(x_1, \ldots, x_d, 1)$$

where $1 \in H_0(B;\mathbf{Z})$. With this definition, we then define the *Donaldson Series*

$$D_B(x) = \sum_{d=0}^{\infty} \frac{\gamma_d(B)(x)}{d!}.$$

While there are no completely general convergence theorems yet proved, in all known examples (including all algebraic surfaces with $b_2^+ > 1$) this series does converge and represents some well-known analytic function. In fact it is now believed[2] to be the case that for any minimal algebraic surface S of general type with $b_2^+(S) > 1$ the Donaldson series of S is

$$D_S(x) = C_0 \exp(q_S/2)\big(\exp(K_S) \pm \exp(-K_S)\big)$$

where K_S is the canonical class of the surface S and C_0 is a constant which depends only on the Euler characteristic of S. There are similar formulas in the case of non-minimal surfaces involving the canonical class of the minimal model and the exceptional curves.

---

[2]Later work with Seiberg-Witten invariants gives extremely strong evidence that this is true

For example when B is the $K3$ we have

$$D_B(x) = \exp\left(q_B(x)/2\right)$$

if B is the blowup of the $K3$ surface at r points we have

$$D_B(x) = \exp\left(q_B/2\right)\left(\sum_{i=1}^{r} \cosh(e_i)\right)$$

where $e_1, \ldots, e_r$ are the exception curves.

Kronheimer-Mrowka [7] have introduced the notion of simple type for a four-manifold B. The condition is that

$$\gamma_d(\alpha_1, \ldots, \alpha, t, 1, 1) = 4\gamma_d(\alpha_1, \ldots, \alpha_t)$$

for all classes $\alpha_i \in H_2(M; \mathbf{Z})$. Under this condition, they go on to show that the Donaldson series D_B is of the form

$$\exp(q_B/2) \sum_i a_i \exp(K_i)$$

for appropriate rational numbers a_i and an appropriate finite set of two-dimensional cohomology classes K_i. For more details on this see Stern's lectures.

7.8. Problems

1. Show that the statement at $D_B(x) = \exp(q_B/2)$ is equivalent to the statement that
$$\gamma_d(B)(x) = \frac{q_B^{d/2}(x)}{2^d d!}$$
for all $d \geq 0$.
2. Prove that for any principal $SO(3)$-bundle ξ we have
$$p_1(\xi) = \mathcal{P}(w_2(\xi)) \pmod 4.$$
3. Let X be an oriented non-compact manifold of dimension n. Show that there is a fundamental locally finite homology class in $H_n^{l.f.}(X; \mathbf{Z})$.
4. Show that $\mathbf{CP}^2$ admits an involution inducing the map (-1) on H^2. Show that this involution can be extended to one on $X \# \mathbf{CP}^2$ which fixes $H^2(X) \hookrightarrow H^2(X \# \mathbf{CP}^2)$ and sends the exceptional class e to $-e$. Does $\mathbf{CP}^2$ admit an orientation reversing diffeomorphism?
5. Let $\mathbf{CP}^2$ be the complex projective plane given its usual complex orientation. Show that if t is even, $\xi_i \in H_2(B)$ for $i = 1, ..., d-t$, $e \in H_2(\mathbf{CP}^2)$, then
$$\gamma_d(B \# \mathbf{CP}^2)(\xi_1, \ldots, \xi_{d-t}, \underbrace{e, \ldots, e}_{t \text{ times}}) = 0.$$
(It is in fact true that $\gamma(B \# \mathbf{CP}^2)$ vanishes identically, but this is requires more machinery, as we shall see in the next lecture.)
6. Let $\overline{\mathbf{CP}}^2$ be the complex projective plane given the orientation opposite to its complex orientation. Show that if t is odd, $\xi \in H_2(B)$, $e \in H_2(\overline{\mathbf{CP}}^2)$, then
$$\gamma_d(B \# \overline{\mathbf{CP}}^2)(\xi_1, \ldots, \xi_{d-t}, \underbrace{e, \ldots, e}_{t \text{ times}}) = 0.$$

(It is not true that $\gamma(B\#\overline{\mathbf{CP}}^2)$ vanishes identically for all B. In fact, there are universal relations expressing the Donaldson polynomials of $B\#\overline{\mathbf{CP}}^2$ in terms of those for B.)

7. Let $Gr_{SU(2)}(\mathbf{C}^n)$ be the Grassmannian of complex 2-planes in $\mathbf{C}^n$ equipped with an $SU(2)$-equivalence class of a special unitary frame. Show that it is a smooth manifold and compute its dimension. Define the tautological principal $SU(2)$-bundle over this manifold. Show that it is a smooth principal $SU(2)$-bundle.
8. Let H be a complex Hilbert space. Define $Gr_{SU(2)}(H)$ in analogy with the finite dimensional grassmannians in the previous problem. Show that $Gr_{SU(2)}(H)$ is a Hilbert manifold, which is infinite dimensional if H is. Identify the tangent space at any point. Define the tautological principal $SU(2)$-bundle over this grassmannian. Show that it is a smooth principal bundle in the category of Hilbert manifolds.
9. Now fix H to be an infinite dimensional separable Hilbert space. We denote by $\zeta \to Gr_{SU(2)}(H)$ the tautological principal $SU(2)$-bundle. We fix a smooth riemannian metric on the tangent bundle to $Gr_{SU(2)}(H)$. Show that if $P \to B$ is a smooth principal bundle, then there is a smooth map $B \to Gr_{SU(2)}(H)$ which is covered by a smooth $SU(2)$-equivariant map $P \to \zeta$. Show all such mapping of B to $Gr_{SU(2)}(H)$ are homotopic.
10. Show that the space of smooth $SU(2)$-equivariant mappings $P \to \zeta$ is contractable. Show that $\mathrm{Aut}(P)$ acts naturally on this space. Show that the action is free and that the quotient space is identified with the component of the space of smooth maps form B to $Gr_{SU(2)}(H)$ consisting of all maps which pull back ζ to a bundle isomorphic to P.
11. Show that the space of $SU(2)$-equivariant mappings of P to ζ which project to L^2_2-mappings from $B \to Gr_{SU(2)}(H)$ is contractable. Show that the group of gauge transformations $\mathcal{G}_3(P)$ acts naturally on this space with the quotient being identified with the space of L^2_2-mappings from B to $Gr_{SU(2)}(H)$. This proves that $Gr_{SU(2)}$ is a classifying space for $\mathcal{G}_3(P)$.
12. Show that $\mathcal{B}^*_2(P)$ is a classifying space for $\mathcal{G}_3(P)/\{\pm 1\}$.
13. Now let H be a separable, infinite dimensional real Hilbert space. Let $Gr_{SO(3)}(H)$ be the grassmannian of oriented 3-planes in H. Show that the space of L^2_2-mappings from B to $Gr_{SO(3)}(H)$ which can be covered by $SO(3)$-equivariant mappings from $P/\{\pm 1\}$ to the total space of the tautological $SO(3)$-bundle is a classifying space for $\mathcal{G}_3(P)/\{\pm 1\}$. Show this space is weakly homotopy equivalent to $\mathcal{B}^*_2(P)$. In particular, this space has the same cohomology ring and homotopy groups as $\mathcal{B}^*_2(P)$.
14. Consider the space of continuous mappings $B \to Gr_{SO(3)}(H)$ which pull back the tautological $SO(3)$-bundle to a bundle isomorphic to $P/\{\pm 1\}$ endowed with the compact-open topology. Show that this space has the same cohomology ring and the same homotopy groups as $\mathcal{B}^*_2(P)$.
15. Compute $\pi_1(\mathcal{B}^*_2(P))$.
16. Compute the rational cohomology ring of $\mathcal{B}^*_2(P)$.
17. Compute $\pi_2(\mathcal{B}^*_2(P))$.

18. Let $P \to B$ be a principal $SU(2)$ bundle over any base B. Let $\mathrm{Ad}(P)$ denote the associated $SO(3)$ bundle. Show that
$$p_1(\mathrm{Ad}(P)) = -\frac{1}{4}c_2(P).$$
19. Show that the "slant product is adjoint to cross-product"; i.e. if $\nu \in H^n(X \times Y)$, $x, y \in H_k(X), H_{n-k}(Y)$ respectively, then
$$\langle \nu/x, y\rangle = \langle \nu, x \times y\rangle.$$
20. Let Y be a three-manifold. Show that a connection A on an $SU(2)$ bundle $P \to Y \times S^1$ induces a closed path $\gamma : S^1 \to \mathcal{A}(N \times SU(2))$, which we can compose with the Chern-Simons map to get a map from $S^1 \to S^1$. Show that the degree of this map is exactly the second Chern class of P.
21. A smoothly embedded three-manifold $N \hookrightarrow B$ induces a map
$$CS_N : \mathcal{B} \to S^1,$$
defined by evaluating the Chern-Simons invariant of a given equivalence class of connections over N. Use Problems 19 and 20 to show that
$$\mu(N) = CS_N^*([S^1]^*),$$
where $[S^1]^*$ is the (appropriate) generator of $H^1(S^1)$.

LECTURE 8
The Connected Sum Theorem

The connected sum theorem and the non-vanishing theorem for algebraic surfaces

In this section we shall state two fundamental theorems of Donaldson and then derive a very general corollary about algebraic surfaces.

8.1. The main results

Let us begin with the two theorems of Donaldson which form the core of this section.

Theorem 8.1.1. (The Connected Sum Theorem) *Let Y and Z be closed, oriented simply connected 4-manifolds with $b_2^+(Y), b_2^+(Z) > 0$. Suppose B is orientation-preserving diffeomorphic to $Y \# Z$. Then $\gamma_d(B) = 0$ for all $d \geq 0$.*

Theorem 8.1.2. Non-vanishing Theorem for Algebraic Surfaces. *Let X be a simply connected algebraic surface. Then for all $d >> 0$, $d \equiv 3(1 + b_2^+(X))/2 \pmod 4$ the Donaldson polynomial invariant $\gamma_d(X)$ is non-trivial.*

Remark 8.1.3. For an algebraic surface X we have $b_2^+(X) = 2p_g(X) + 1$, where $p_g(X)$ is the geometric genus and in particular is an integer. Thus, $(1+b_2^+(X))/2 = p_g(X) + 1$ and d is an integer.

Let us establish some corollaries of these two theorems.

Corollary 8.1.4. *If X is the $K3$ surface and if X is orientation-preserving diffeomorphic to $Y \# Z$, then one of Y and Z is a 4-homotopy sphere.*

Proof. Suppose that X is orientation-preserving diffeomorphic to $Y \# Z$. Of course, since X is simply connected, so are Y and Z. Since $\gamma_d(X) \neq 0$ for all $d >> 0$, we see that one of Y and Z must be negative definite. By symmetry we can assume that it is Y. According to Donaldson's theorem about definite manifolds, the intersection form of Y must be equivalent over $\mathbf{Z}$ to the diagonal form with -1's down the diagonal. Since the intersection form of X and therefore of Y is even, it can have no classes of square -1. The only way to reconcile these two facts is that $H^2(Y; \mathbf{Z})$ be zero. That is to say Y is a homotopy 4-sphere. □

This answers in a very strong fashion the question asked in an earlier section about the $K3$ surface – there is no decomposition of a $K3$ surface as $(S^2 \times S^2)\#Z$.

The only properties that we used about the $K3$-surface in the above were that it was a simply connected algebraic surface and that it was a spin manifold (i.e., the intersection form is even). Thus, we have a vast generalization of this result.

Corollary 8.1.5. *Let X be a simply connected algebraic surface with even intersection form. If X is orientation-preserving diffeomorphic to $Y\#Z$ then one of Y and Z is a homotopy 4-sphere.*

One might ask if this result can possibly hold without the assumption that the intersection form is even. It is easy to see that if X is a simply connected algebraic surface and if X' denotes the blow up of X at one point, then X' is a simply connected algebraic surface and X' is orientation-preserving diffeomorphic to $X\#\overline{\mathbf{CP}}^2$. Of course if we blow up k points then we obtain a surface diffeomorphic to $X\#_k\overline{\mathbf{CP}}^2$ Essentially, this is the only way that algebraic surfaces can decompose.

Corollary 8.1.6. *Suppose that X is a simply connected algebraic surface and that X is orientation-preserving diffeomorphic to $Y\#Z$. Then one of Y and Z is homotopy equivalent to $\#_k\overline{\mathbf{CP}}^2$ for some $k \geq 0$.*

Remark 8.1.7. In all known examples, if an algebraic surface X is orientation-diffeomorphic to $Y\#Z$ with the intersection form of Z negative definite then Y is diffeomorphic to an algebraic surface and the complex structure on X can be deformed to the complex structure on Y blown up at k points.

It follows from the classification of odd, indefinite, unimodular pairings that a simply connected algebraic surface X with an indefinite odd intersection form is homotopy equivalent to a connected sum

$$\#_k\mathbf{CP}^2\#_\ell\overline{\mathbf{CP}}^2.$$

As long as $b_2^+(X) > 1$ it follows from the two Donaldson theorems that X is not diffeomorphic to $\#_k\mathbf{CP}^2\#_\ell\overline{\mathbf{CP}}^2$. This gives us examples of two homotopy equivalent, non-diffeomorphic 4-manifolds. In fact, one can find infinitely many pairwise non-diffeomorphic manifolds which are all homotopy equivalent. These manifolds are algebraic surfaces and are distinguished by their Donaldson polynomial invariants.

8.2. The divisors in $\mathcal{B}_2^*(Q)$ representing $\mu(x)$

In this subsection and the next one we outline the proof of the connected sum theorem.

Let us fix a simply connected 4-manifold B with $b_2^+(B) > 1$. We fix a generic metric on B. For each $r \geq 0$ we denote by $P_r \to B$ a principal $SU(2)$-bundle with $c_2 = r$. Let $x \in H_2(B;\mathbf{Z})$ be a homology class. Represent x by an embedded surface $\Sigma \subset B$. Then restricting connections gives us a map

$$\mathcal{B}_2(P_r) \to \mathcal{B}_1(P_r|_\Sigma)$$

from the space of L_2^2-connections for P_r to the space of L_1^2-connections for $P_r|_\Sigma$. (Because we are restricting from a four-manifold to a two-manifold we lose control on one derivative in the Sobolev norms.) Of course, all principal $SU(2)$-bundles over Σ are isomorphic. Thus we can identify $P_r|_\Sigma$ with the trivial bundle $\Sigma \times SU(2)$. We can restrict to the moduli spaces to produce

$$\mathcal{M}(P_r) \to \mathcal{B}_1(\Sigma \times SU(2)).$$

Because we have chosen a generic metric on B and since $b_2^+(B) > 0$, if $r > 0$, then every ASD connection on P_r is irreducible. It is a technical fact that the restriction of these connections to Σ must also be irreducible, at least if we choose the embedding of Σ in B generically. Thus, for every $r > 0$ restriction induces a map

$$\mathcal{M}(P_r) \to \mathcal{B}_1^*(\Sigma \times SU(2)).$$

Just as in the case of 4-manifold bases there is a universal $SO(3)$-bundle over $\Sigma \times \mathcal{B}_1^*(\Sigma \times SU(2))$. Slanting with one-quarter the first Pontrjagin class of this bundle defines a map

$$\mu_\Sigma \colon H_2(\Sigma; \mathbf{Z}) \to H^2(\mathcal{B}_1^*(\Sigma \times SU(2)); \mathbf{Z}).$$

The pullback of these classes under the restriction mappings give the classes $\mu([\Sigma]) \in H^2(\mathcal{M}(P_r); \mathbf{Z})$. Let us choose a divisor representative $D_\Sigma \subset \mathcal{B}_1^*(\Sigma \times SU(2))$ for $\mu_\Sigma([\Sigma]) \in H^2(\mathcal{B}_1^*(\Sigma \times SU(2)); \mathbf{Z})$. If we choose these divisors generically then the maps from the moduli spaces $\mathcal{M}(P_r)$, $r > 0$, will be transverse to these divisors, and hence the preimages of the divisor D_Σ under the restriction mapping will give divisors in the moduli spaces $\mathcal{M}(P_r)$ representing $\mu([\Sigma])$.

Such a divisor in $\mathcal{M}(P_r)$ representing $\mu([x])$ has the property that its locus is completely determined by the restriction of the connections to $\Sigma \subset B$.

We have the following corollary for representative divisors of this type. This corollary will play an important role in what follows.

Proposition 8.2.1. *Suppose that $D_k \in \mathcal{M}(P_k)$ is a divisor representing $\mu(x)$ and induced from a universal divisor $D_\Sigma \subset \mathcal{B}_1^*(\Sigma \times SU(2))$ for some surface Σ representing x. Suppose that $[A_n] \in \mathcal{M}(P_k)$ is a sequence of points contained in D_k. Suppose that this sequence has limit $([A], \{x_1, \ldots, x_t\})$ in the Uhlenbeck compactification $\overline{\mathcal{M}}(P)$. Then either some x_i lies on Σ or $[A]$ lies in the divisor $D_{k-t} \subset \mathcal{M}(P_{k-t})$ induced by restriction from the universal divisor D_Σ.*

Proof. Suppose that no x_i lies on Σ. Then the restrictions $[A_n|_\Sigma]$ converge in $\mathcal{B}_1^*(\Sigma \times SU(2))$ to $[A|_\Sigma]$. Since the divisor $D_\Sigma \subset \mathcal{B}_1^*(\Sigma \times SU(2))$ is a closed subset and since $[A_n|_\Sigma] \in D_\Sigma$, it follows that $[A|_\Sigma] \in D_\Sigma$. That is to say, $[A]$ lies in the divisor in $\mathcal{M}(P_{k-t})$ induced by restriction from D_Σ. □

Remark 8.2.2. In fact, in the intersection of $\mathcal{M}(P)$ with a small neighborhood of $\mathcal{M}(P_{k-r}) \times \Sigma^r(B)$ in $\overline{\mathcal{M}}(P_k)$, the divisor D_k is close to $D_{k-r} \otimes 1 + 1 \otimes \mathcal{I}(\Sigma)$, where $\mathcal{I}(\Sigma) \hookrightarrow \Sigma^r(B)$ is the divisor of all r-tuples of points, at least one of which lies in Σ. We can also choose D_Σ so that its support is disjoint from connections on Σ which are close to the product connection. Hence, in the intersection of $\mathcal{M}(P_k)$ with a neighborhood of $\mathcal{M}(P_0) \times \Sigma^k(B)$, D_k is close to $1 \otimes \mathcal{I}(\Sigma)$.

8.3. The Taubes gluing setup of the connected sum theorem

The result we give here is a consequence of the ideas used in the proof of the Taubes gluing theorem. We choose metrics on X and Y and small balls D and D' about the points $x \in X$ and $y \in Y$ at which we shall do the connected sum and fix an orientation-reversing isometry $TX_x \to TY_y$. We fix an annulus along which to invert. This annulus is chosen to have a sufficiently large conformal radius. We now choose a metric on the connected sum which is equal to the restrictions of the metrics on X and Y away from the gluing annulus and which is close to a long cylindrical metric $S^3 \times [0,T]$ on the gluing region.

We call this manifold B. Fix k and let $Q \to B$ be a principal $SU(2)$-bundle with $c_2(Q) = k$.

There are constants $L > 2$ and $\epsilon > 0$ such that any ASD connection on the cylinder $S^3 \times [0,L]$ with $\|F_A\|_{L^2} < \epsilon$ is L^2_2-close to a product connection when restricted to $S^3 \times [1, L-1]$.

If T is sufficiently large, then we can choose subannuli $A_1, \dots, A_t$ inside $S^3 \times [0,T]$ such that each annulus is of length L and any ASD connection has energy less than $\epsilon > 0$ on at least one of the A_i.

For each annulus A_i, there are corresponding disks $D_i \subset D \subset X$ and $D_i' \subset D' \subset Y$. We choose these disks so that the following hold. The disks are centered at x and y respectively and have boundary which agree with a boundary component of A_i, and $A_i \subset D_i$ and $A_i' \subset D_i'$. For each $r > 0$ we let $P_r \to X$ and $P_r' \to Y$ be principal $SU(2)$-bundles with Chern class r.

We define open subsets V_1 and V_2 of $\overline{\mathcal{M}}(P)$ as follows. A generalized connection $[A]$ is contained in V_1 if there is a finite set of disjoint balls $B_1, \dots, B_s$ of small radius such that the integral of the measure of A over $X - D_0 \cup B_i$ with respect to this connection is small and the integral of the measure of A over each of the balls is approximately equal to $8\pi^2 n_i$ for some integer $n_i > 0$. We define V_2 similarly, using Y instead of X. We have similarly-defined open subsets of $\mathcal{M}(P_r)$ and $\mathcal{M}(P_r')$ denoted by U and U' respectively.

Proposition 8.3.1. *For each i and any $0 < r, k$, let $W_i^x(P_r)$ be the subspace of the based moduli space ${}^x\mathcal{M}^*(P_r)$ consisting of based connections whose energy is less than ϵ on D_i and which do not project to points in the closure of U. We define $W_i^{x'}(P_r')$ similarly. The gluing construction defines a diffeomorphism*

$$W_i^x(P_r) \times_{SO(3)} W_i^y(P_{k-r}') \to \mathcal{M}(Q)$$

onto an open subsets. The connections in the image of this gluing map have small curvature over the annulus A_i. The union over all i and all $0 < r < k$ of the images of these gluing maps contains all the connections in $\mathcal{M}(Q)$ which are not contained in the closure of either V_1 or V_2.

Suppose that Σ is a surface embedded in $(X - D)\coprod(Y - D')$. Then if we glue up ASD connections A and A' on X and Y we produce a connection $\tilde{A}\#\tilde{A}'$ which agrees with A over $(X - D)\coprod(Y - D')$ and in particular over Σ. When we apply the Taubes gluing construction the connection $\tilde{A}\#\tilde{A}'$ is deformed to an ASD connection $T(A, A') = (\tilde{A}\#\tilde{A}') + \alpha$ where α has a small L^2_2-norm. Furthermore, the estimate on the norm of α goes to zero as the conformal length L of the annuli goes to infinity. Thus, we see that the restrictions over Σ of $A\coprod A'$ and $T(A, A')$ are L^2_2-close, and the estimate of how close goes to zero as L goes to infinity. In fact, not only are the restrictions to Σ L^2_2-close, the differentials of the restriction mappings are also close.

Since the maps

$$W_i^x(P_r) \to \mathcal{B}_1^*(\Sigma \times SU(2))$$

are transverse to D_Σ, it follows that if L is sufficiently large, it will be the case that the map

$$\overline{\mathcal{M}}(P) - (V_1 \cup V_2) \to \mathcal{B}_1^*(\Sigma \times SU(2))$$

has image contained in the irreducible connections, and the map is transverse to D_Σ. The preimage of this divisor is a divisor representing the restriction of $\mu([\Sigma])$ to this open subset.

The induced divisor completes over $V_1 \cup V_2$, giving a representative on all of $\overline{\mathcal{M}}(P)$ for the class $\mu([\Sigma])$. If $\Sigma \subset X$, then the restriction of the extension of the divisor to V_1 is close to the subset of points which have a point of singular curvature near a point on Σ. The restriction of the extension of the divisor to $V_2 \setminus V_1$ is pulled back from D_Σ by the restriction mapping. The analogous statements hold when $\Sigma \subset Y$.

8.4. The sketch of the argument for the connected sum theorem

We are ready to complete the argument for the connected sum theorem. Let B be orientation-preserving diffeomorphic to $X\#Y$, with $b_2^+(X), b_2^+(Y) > 0$. Choose generic metrics on X and Y so that the only reducible ASD connections over principal $SU(2)$-bundles on these manifolds are product connections. Let us suppose that we have classes $x_1, \ldots, x_d \in H_2(X;\mathbf{Z}) \oplus H_2(Y;\mathbf{Z})$. We shall only consider the special case when the number of classes contained in X, say a, is at least $3(1 + b_2^+(X))/2$, and similarly the number of classes contained in Y, say b, is at least $3(1 + b_2^+(Y))/2$. Under these conditions on the number of classes in each of X and Y we shall prove that the value of the Donaldson polynomial vanishes on the given collection of homology classes. The other cases can be handled either by appealing to a blow-up argument or by a direct argument which uses a different circle of ideas from the ones we introduce here.

We represent the classes $x_1, \ldots, x_d$ by generically embedded surfaces Σ_i in $X - D \coprod Y - D'$. We assume that these surfaces meet transversely. By pulling back universal divisors D_{Σ_i} and extending as described above we obtain divisors in $\overline{\mathcal{M}}(P)$ representing the images under μ of the x_i.

If the value of the Donaldson polynomial on these classes is non-zero, then the divisors representing these classes must have a non-empty intersection.

Lemma 8.4.1. *If a point of V_1 lies in the intersection of the all the divisors representing the classes $\mu(\Sigma_i)$ for $\Sigma_i \subset X - D$, then the charge on $X - D$ of the generalized connection representing this point is at least $a/2$. (Similarly for Y.)*

Proof. In this region of the moduli space the divisors associated to $\mu(\Sigma_i)$ are close to the subset of points with a singular mass along Σ_i. Since there are a of the Σ_i and since no three of the Σ_i meet at any point of X, it follows that any point of intersection of the a divisors must lie near a point represented by a generalized connection whose measure has at least $a/2$ points of singular support in $X - D$. This means that the integral of the measure over $X - D$ is at least $8\pi^2(a/2)$, i.e., the charge on X of this generalized connection is at least $a/2$. □

Now we are ready to consider the possibilities. First, we examine whether there can be a point of intersection in $V_1 \cap V_2$. Suppose that there is such a point and that the distribution of its charge is r on $X - D$ and s on $Y - D$. If so then we have inequalities $2a \leq 4r$ and $2b \leq 4s$. This implies that $2a + 2b \leq 4(r+s)$. Of course, we have

$$2a + 2b = 8(r+s) - 3(1 + b_2^+(X\#Y)).$$

Thus, we have

$$2a + 2b \leq 4r + 4s \leq 3(1 + b_2^+(X\#Y)) < 3(1 + b_2^+(X)) + 3(1 + b_2^+(Y)).$$

This contradicts our assumption on a and b.

Next, we consider the case of a point of intersection in $V_1 \setminus V_2$. Again let us label the charge distribution of a point of intersection of the divisors as r on X and s on Y. In this case we have

$$2a \leq 4r$$

and

$$2b \leq 8s - 3(1 + b_2^+(Y)).$$

Since $2a \geq 3(1 + b_2^+(X))$ and $2a + 2b = 8(r+s) - 3\big(1 + b_2^+(X\#Y)\big)$, we conclude that

$$8(r+s) - 3(1 + b_2^+(X\#Y)) \leq 4r + 8s - 3(1 + b_2^+(Y)),$$

which in turn implies that

$$2a \leq 4r < 3(1 + b_2^+(X)),$$

contradicting our assumption on a. Symmetry handles the case of a point of intersection in $V_2 \setminus V_1$.

Lastly, we examine the possibility that there is a point of intersection outside on $V_1 \cup V_2$. Such a point of intersection will lie in one of the Taubes gluing neighborhoods. Thus, we will have some $0 < r < k$ and some i for which the embedding

$$W_i^x(P_r) \times_{SO(3)} W_i^{x'}(P'_{k-r}) \subset \overline{\mathcal{M}}(P)$$

containing the point of intersection. But the intersection of the divisors representing $\mu(x_i)$ for Σ_i in X are close to the pullbacks of the corresponding divisors from $W_i(P_r)$. This means that the intersection of these divisors in $\mathcal{M}(P_r)$ must be non-empty, which in turn implies that $2a \leq 8r - 3(1 + b_2^+(X))$. Similarly, we have $2b \leq 8(k-r) - 3(1 + b_2^+(Y))$. Putting these together gives us $2a + 2b \leq 8k - 3(2 + b_2^+(X\#Y))$, which is absurd since $2a + 2b = 8k - 3(1 + b_2^+(X\#Y))$.

This completes the sketch of the proof of the connected sum theorem.

8.5. The non-vanishing theorem

The proof of the non-vanishing theorem uses a relationship between the anti-self-dual connections and holomorphic vector bundles. The basic result is that if $P \to S$ is a principal $SU(2)$-bundle over an algebraic surface S, then a connection on P which is ASD with respect to a Kähler metric on S determines a holomorphic bundle structure for the vector bundle $P \times_{SU(2)} \mathbf{C}^2 \to S$. (See Jun Li's lectures.) Furthermore, if A is irreducible then this bundle has a special property – it is stable with respect to the Kähler metric. Donaldson proved the converse, every stable holomorphic bundle $\mathbf{C}^2$ with trivial determinant bundle over an algebraic surface admits a hermitian metric and an $SU(2)$-connection for this metric which is both a holomorphic connection and is ASD with respect to the Kähler metric. This connection is irreducible. It is easy to see that this sets up an isomorphism between the moduli space $\mathcal{M}(P)$ and the moduli space of stable holomorphic bundles over S. For more details see Jun Li's lectures.

We have the Uhlenbeck compactification $\overline{\mathcal{M}}(P)$ of the space of ASD connections. There is also a compactification, the Gieseker compactification, of the moduli space of holomorphic bundles over S. It is a projective variety. The diffeomorphism between these moduli spaces extends to a degree one map from the Gieseker compactification of the moduli space of stable bundles to the Uhlenbeck compactification $\overline{\mathcal{M}}(P)$. There is an algebro-geometric version of the μ map, denoted ν from $H_2(S;\mathbf{Z})$ to the second integral cohomology of the Gieseker compactification. This map is constructed using a tautological sheaf over the product of S with the Gieseker compactification. (This sheaf is the analogue of the universal $SO(3)$-bundle in gauge theory.) One shows that the pullback of the μ map from $H^2(\overline{\mathcal{M}}(P);\mathbf{Z})$ to the integral cohomology of the Gieseker compactification is ν. Furthermore, if we apply ν to the homology class represented by a hyperplane section of $H \subset S$, then we obtain the cohomology class of an ample divisor on the Gieseker compactification. Thus, provided that the Gieseker moduli space is of the correct dimension (i.e., if its geometric dimension is equal to its formal dimension), the value of the ν-map on d copies of the hyperplane section will be positive. It follows from the fact that the map from the Gieseker compactification to $\overline{\mathcal{M}}(P)$ is degree one, that the value of the Donaldson polynomial on d copies of $[H]$ is also positive, and hence the Donaldson polynomial is non-trivial.

It remains to show that the Gieseker moduli space is of the correct dimension. By deformation theory we know that the Gieseker moduli space is everywhere of at least its formal dimension. Any point where the geometric dimension is larger than the formal dimension is a singular point. Thus, if the geometric dimension of the variety is not everywhere equal to its formal dimension, then the geometric dimension of the singular set is greater than the formal dimension of the variety. Donaldson showed that provided that the Chern class c_2 was sufficiently large, the dimension of the singularities of the Gieseker moduli space was less than the formal dimension of this variety. It then follows that the variety is generically smooth with geometric dimension equal to its formal dimension.

8.6. Problems

1. Show that if $a \in H^2(M,\mathbf{Z})$, with M an oriented smooth manifold, then there is a oriented, codimension-two smooth submanifold $D \hookrightarrow M$ whose fundamental class is Poincaré dual to a.
2. Show that if X is a triangulated space, $a \in H^2(X,\mathbf{Z})$, then there is a sequence of inclusions
$$D \hookrightarrow \xi \hookrightarrow X,$$
where ξ is an oriented real two-plane bundle over D, the inclusion of D into ξ is the zero-section, and the total space of ξ is embedded as an open subset of X so that a is the image in $H^2(X)$ of the Thom class $U \in H^2(\xi, \xi - D)$.
3. Show every principal $SU(2)$ bundle over a two-manifold Σ is trivial.
4. Define the universal $SO(3)$ bundle over $\Sigma \times \mathcal{B}_1^*(\Sigma \times SU(2))$.
5. Show that the conditions in Section 8.5 imply that any point of intersection of all the divisors associated with the $\Sigma_i \hookrightarrow X$ has energy over X in the stable range.

6. Show that if $f\colon X \to Y$ is a map of non-zero degree between compact, oriented manifolds and if $x_1, \ldots .x_t$ are cohomology classes in Y with the property that the sum of their degrees is equal to the dimension of Y, show that $x_1 \cup \cdots \cup x_t \neq 0$ if and only if $f^*x_1 \cup \cdots \cup f^*x_t \neq 0$.

References

[1] M. Atiyah, *Geometry of Yang-Mills Fields*, Academia Nazionale Dei Linchi Scuola Normale Superiore, Lezioni Fermiane, Pisa 1979.

[2] S. Donaldson, *The Orientation of Yang Mills moduli spaces and four manifold topology*, J. Diff. Geo. **26** (1987) 397-428.

[3] S. Donaldson and P. Kronheimer, *The Geometry of Four-Manifolds*, Clarendon, Oxford 1990.

[4] D. Freed and K. Uhlenbeck, *Instantons and Four-manifolds*, Mathematical Sciences Research Institute Publications 1. Springer, New York 1984.

[5] R. Friedman and J. Morgan, *Smooth Four-Manifolds and Complex Surfaces*, Ergebnisse der Mathematik und ihrer Grenzgebiete 3 Folge Band 27, Springer-Verlag, Berlin, Heidelberg 1994.

[6] D. Kotschick and J. Morgan, *$SO(3)$-invariants for 4-manifolds with $b_2^+ = 1$*, Part II, J. Diff. Geo. **39** (1994) 433-456.

[7] P. Kronheimer and T. Mrowka, Embedded Surfaces and the Structures of Donaldson's Polynomial Invariants, J. Diff. Geo **41** (1995) 573-735.

[8] R. Palais, *Foundations of Global Analysis*, Benjamin, New York, 1968.

[9] C. Taubes, Self-dual connections on 4-manifolds with indefinite intersection matrix, J. Diff. Geo. **19** (1984) 517-560.

Computing Donaldson Invariants

Ronald J. Stern

IAS/Park City Mathematics Series
Volume 4, 1998

Computing Donaldson Invariants

R. J. Stern

Abstract. In these lectures we will discuss techniques for constructing simply-connected smooth 4-manifolds and computing their Donaldson invariants. There are 37 Exercises (which can be solved by the diligent reader) and 24 Problems whose solutions should advance our understanding of smooth 4-manifolds.[4] Most of the material in these lectures is either drawn from (or is the basis of) other work by the author and Ronald A. Fintushel.

[1]Department of Mathematics, University of California Irvine, California 92717.
E-mail address: `rstern@math.uci.edu`.

[2]The author was supported in part by NSF Grant DMS9302526.

[3]*1991 Mathematics Subject Classification.* Primary 57R57, 57N13, 57R55; Secondary 58D29, 53C20.

[4]Added in proof: These lectures were delivered in July 1994. A few months later, gauge theory underwent the Seiberg-Witten revolution. Many of the results presented in these lectures now have more easily stated and proved counterparts in this new Seiberg-Witten theory. However, of the 24 Problems presented in these lectures, only Problems 5, 7, 8, 11, 12, 16 have either partial or complete answers. These new developments (up to June 1997) will appear as footnotes.

LECTURE 1
Overview

1.1. Classical invariants

Our problem is to classify closed smooth 4-manifolds. To avoid the group theoretic problems arising from the fact that any finitely presented group can occur as the fundamental group of a smooth closed 4-manifold, we assume our manifolds are simply-connected. Most of the classical invariants for 4-manifolds are encoded by the intersection form Q_X. This form is an integral unimodular symmetric bilinear pairing

$$Q_X : H_2(X;\mathbf{Z}) \otimes H_2(X;\mathbf{Z}) \to \mathbf{Z}$$

obtained by representing homology classes as oriented embedded submanifolds and counting intersections with signs; it is Poincaré dual to the pairing given by cup product. From the intersection form one can determine its **rank** (which is the second Betti number $b_2(X) = \text{rank} H_2(X;\mathbf{Z}) = b^+(X) + b^-(X)$), its **signature** $= \sigma(X) = b^+(X) - b^-(X)$ (where $b^{\pm}(X)$ are the dimensions of the $\pm$-eigenspaces of Q_X) and its **type** (which is **even** if $Q_X(x,x) \equiv 0 \mod 2$ for all x, and **odd** otherwise). A form is **definite** provided one of $b^{\pm}(X)$ vanishes, and is **indefinite** otherwise. Note that the Euler characteristic $e(X)$ of X is $b^+(X)+b^-(X)+2$. Also, a simply-connected 4-manifold with even intersection form is a spin 4-manifold (i.e. $w_2(X) = 0$). Further, $Q_{X\#Y} = Q_X \oplus Q_Y$. Given a basis for $H_2(X;\mathbf{Z})$ we can represent Q_X as a symmetric integral matrix with determinant equal to ± 1.

The standard examples of 4-manifolds and their intersection forms are:

S^4	$Q_{S^4} = \emptyset$
CP^2	$Q_{\mathrm{CP}^2} = (1)$
$S^2 \times S^2$	$Q_{S^2\times S^2} = \begin{pmatrix} 0 & 1 \\ 1 & 0 \end{pmatrix} = H$
K3	$Q_{K3} = E_8 \oplus E_8 \oplus 3H$

Here $K3$ is the Kummer or K3 surface which is diffeomorphic to the quartic surface in CP^3 and E_8 is the intersection form for the E_8 plumbing manifold:

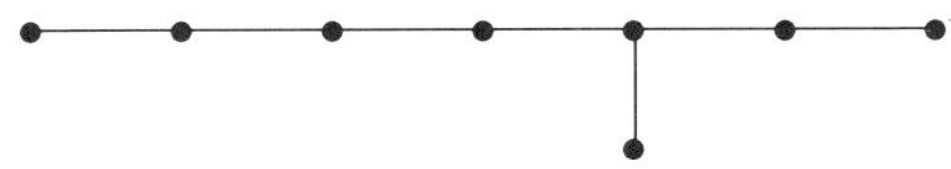

Here, each node denotes the unit cotangent bundle of S^2; an interval indicates that the endpoint disk bundles are plumbed, i.e identified fiber to base over the upper hemisphere of each S^2. Then each of the eight nodes represents an S^2 with self-intersection -2 so that E_8 is a 4-manifold with negative definite intersection form and with boundary the Poincare' homology 3-sphere.

The intersection form is a fundamental homotopy invariant.

1.1.1. *Whitehead's Theorem* [Wh]. *The homotopy type of an oriented simply connected 4-manifold is completely determined by its intersection form.*

The fundamental work of M. Freedman establishes the intersection form as a fundamental topological invariant.

1.1.2. *Freedman's Theorem* [F]. *The homeomorphism type of an oriented simply connected smooth 4-manifold is completely determined by its intersection form.*

Apriori, the signature of any even integral symmetric unimodular bilinear form is divisible by 8. An important classical theorem that provided much of the vitality for 4-manifolds prior to Freedman-Donaldson is:

1.1.3. *Rohlin's Theorem* [R]. *The signature (of the intersection form) of a smooth spin 4-manifold is divisible by 16.*

Although, as we shall see, the smooth topology of a 4-manifold is rather delicate, Wall's diffeomorphism theory says things are stably as nice as can be expected. For example:

1.1.4. *Wall's Stability Theorem* [W2, W3]. *If two simply-connected closed smooth 4-manifolds have isomorphic intersection forms, then they are "stably diffeomorphic", that is, they become diffeomorphic after connect summing with a number of copies of $S^2 \times S^2$. (This number is unspecified.)*

1.1.5. *Wall's Diffeomorphism Theorem* [W2, W3]. *If X is a simply connected smooth 4-manifold with Q_X indefinite, then the homology induced map*

$$\mathit{Diff}(X\#S^2 \times S^2) \to O < H_2(X\#S^2 \times S^2; \mathbf{Z}), Q_X >$$

is an isomorphism.

Given two homeomorphic simply-connected closed smooth 4-manifolds X and Y, it is then an interesting problem to determine the minimal integer $k(X, Y)$ for which $X\#_{k(X,Y)}S^2 \times S^2$ is diffeomorphic to $Y\#_{k(X,Y)}S^2 \times S^2$.

Problem 1. Determine $k(X, Y)$; despite some heavy input from gauge theory ([**D7**],[**FS5**]) it is still reasonable to conjecture that $k(X, Y) \leq 1$.

The fundamental questions concerning 4-manifolds still remain.

1.1.6. *Existence.* *Determine the unimodular, symmetric, bilinear integral forms that are realized as the intersection form of a closed, simply-connected smooth 4-manifold.*

1.1.7. *Uniqueness.* *Given a closed, simply-connected smooth 4-manifold M, determine the distinct smooth structures on M.*

These questions are the subject of active current research.

1.2. Existence

The early work of Donaldson and its derivatives ([**D1**],[**D3**],[**FS1**],[**FS2**]) has shown that there are indeed restrictions on the intersection form of a closed, simply-connected 4-manifold and has made progress towards a verification of the following existence conjecture:

1.2.1. *11/8 Conjecture.* *The intersection form of a closed, simply-connected smooth 4-manifold must be either*

- *diagonalizable (over $\mathbf{Z}$), or*
- *even and $\frac{b_2(X)}{|\sigma(X)|} \geq \frac{11}{8}$.*

Exercise 1. Complex surfaces and their connected sums satisfy the 11/8 Conjecture.

All diagonalizable forms are realized by the connected sums of $\pm CP^2$. Although the classification of definite integral forms is an active area of research and far from an accomplished feat, nature has a way of taking care of all this. Donaldson [**D1**] first showed that no non-diagonalizable definite form can be realized as the intersection form of a closed smooth 4-manifold; shortly thereafter he [**D3**] showed that if the form $nE_8 \oplus mH$, $n \neq 0$, is realized, then $m \geq 3$. The $K3$-surface realizes $m = 3$.

There has not been much progress on the 11/8 Conjecture beyond the original work of Donaldson. One approach to these existence questions amounts to constructing 4-manifolds with a given Euler characteristic, signature, and type; i.e. it is a geography question. For a long time the only manifolds around were complex manifolds and their connected sums. A long standing conjecture held that this was all there was; i.e., every smooth simply connected closed 4-manifold is the connected sum of manifolds, each admitting a complex structure with one of its orientations. (For this conjecture, the 4-sphere was considered as the empty connected sum.) In 1990 this was disproved by Gompf and Mrowka [**GM**] who produced infinite families of examples of 4-manifolds homeomorphic to the K3 surface and which admit no complex structure with either orientation. Then in 1991 Fintushel and I [**FS6**] constructed irreducible 4-manifolds that (with either orientation) were not homeomorphic to any complex surface. Here, by an *irreducible* 4-manifold we mean that for each splitting as a connected sum $X_1 \# X_2$, one of the X_i is a homotopy sphere (thus avoiding the 4-dimensional Poincaré conjecture). In these lectures we will give more elementary constructions of such manifolds. This work will suggest an even stronger conjecture.

1.2.2. *3/2 Conjecture.* *The intersection form of an irreducible simply-connected closed smooth 4-manifold has the property that $2e(X) - 3|\sigma(X)| \geq 0$.*

Figure 1 illustrates those regions for which irreducible simply-connected closed, smooth 4-manifolds are known to exist. The co-ordinates mimic those for the corresponding geography question for complex surfaces; for a complex surface X, $c_1^2(X) = 2e(X) + 3\sigma(X)$ and its holomorphic Euler characteristic $\chi(X)$ is given by $\chi(X) = (b^+(X)+1)/2 = \frac{1}{12}(c_1^2 + c_2)$ which, being an integer, implies $b^+(X)$ is odd. Further, for minimal surfaces of general type the Noether inequality states that $c_1^2(X) \geq 2\chi - 6$ and the Bogomolov-Miyaoka-Yau inequality states that $c_1^2(X) \leq 9\chi$.

Exercise 2. If $c_1^2(X) \geq 8\chi$, then the signature of X is positive.

From this picture we see that the elliptic surfaces without multiple fibers $E(n)$ (which are irreducible, provided $n > 1$, simply-connected smooth 4-manifolds with signature $= \sigma = -8n$, Euler characteristic=e=$12n$, and type equivalent to n mod 2) are the "largest" known simply-connected irreducible 4-manifolds with a given b^+. An important problem is

Problem 2. Determine canonical procedures for producing irreducible simply-connected smooth 4-manifolds.

For surfaces of general type (which populate the region $2\chi - 6 \leq c_1^2 \leq 9\chi$) it is known that if $2\chi - 6 \leq c_1^2 \leq 8\chi$ then any (c_1^2, χ) is realized by simply-connected genus two Lefschetz fibration. Further there are simply-connected examples provided $\frac{352}{89}\chi + 140.2\chi^{2/3} < c_1^2 < \frac{18644}{2129}\chi - 365.7\chi^{2/3}$. (cf. [**C**])

Exercise 3. For an almost complex structure on a 4-manifold $c_1^2 + c_2 \equiv 0$ (12), i.e $\sigma + e \equiv 0 \mod (4)$. Hence, with the orientation determined by this almost complex structure, if X is simply-connected, $b^+(X)$ is odd. (Recall that an almost complex structure is an endomorphism $J : TX \to TX$ with $J^2 = -\mathrm{Id}$.)

Exercise 4. An oriented simply-connected 4-manifold X has an almost complex structure if and only if $b^+(X)$ is odd.

Exercise 5. The tangent bundle of a simply-connected 4-manifold X splits as the Whitney sum of two complex line bundles if and only if both $b^+(X)$ and $b^-(X)$ are odd.

Problem 3. Does every irreducible simply-connected 4-manifold have an almost complex structure?

All known (simply-connected) irreducible 4-manifolds have, with one orientation, b^+ odd.

Problem 4. Determine the existence of an irreducible simply-connected smooth spin 4-manifold with even $b^+ > 0$; determine the existence of one such manifold with more than one smooth structure, e.g. examples of two simply connected homeomorphic 4-manifolds X and Y with odd b^+ such that $X \# S^2 \times S^2$ is not diffeomorphic to $Y \# S^2 \times S^2$. (cf. Problem 1, [**FS5**]).

Problem 5. Determine whether every irreducible smooth simply-connected 4-manifold with $b^+ > 1$ has a symplectic structure. [1] Conversely, determine if every minimal symplectic 4-manifold is irreducible. [2]

---

[1]Added in proof. There are now many examples of irreducible simply-connected smooth 4-manifolds with no symplectic structures. C.f. Fintushel and Stern, Knots, Links, and 4-manifolds, dg-ga/9702003 and Z. Szabo, Simply connected irreducible 4-manifolds with no symplectic structure, preprint.

[2]Added in proof. True. See Taubes, The Seiberg-Witten and Gromov Invariants, MRL (1995), 221-238.

Problem 6. Find an example of a smooth simply-connected 4-manifold with $b^+ > 1$ with at most finitely many smooth structures.

1.3. Uniqueness (The Donaldson Invariant)

No reasonable Uniqueness Conjecture has emerged. (However unreasonable ones will emerge during these lectures.) Herein lies the current active work in this area. Here, we are studying smooth 4-manifolds which have the same intersection form (and hence are homeomorphic).

Exercise 6. The intersection form of a simply-connected smooth 4-manifold is completely determined by its rank, signature, and type.

Thus a new invariant is required and this begins another of Donaldson's contributions. Given an oriented simply connected 4-manifold with a generic Riemannian metric and an $SU(2)$ or $SO(3)$ bundle P over X, the moduli space of gauge equivalence classes of anti-self-dual connections on P is a manifold $\mathcal{M}_X(P)$ of dimension

$$8\,c_2(P) - 3\,(1 + b_X^+)$$

if P is an $SU(2)$ bundle, and

$$-2p_1(P) - 3\,(1 + b_X^+)$$

if P is an $SO(3)$ bundle. It will often be convenient to treat these two cases together by identifying $\mathcal{M}_X(P)$ and $\mathcal{M}_X(\mathrm{ad}(P))$ for an $SU(2)$ bundle P. Over the product $\mathcal{M}_X(P) \times X$ there is a universal $SO(3)$ bundle $\mathbf{P}$ and there results a homomorphism $\mu : H_i(X) \to H^{4-i}(\mathcal{M}_X(P))$ obtained by decomposing the class $-\frac{1}{4}p_1(\mathbf{P}) \in H^4(\mathcal{M}_X \times X)$. (Homology is always taken to have real coefficients unless it is otherwise adorned.) The basic idea of Donaldson's theory is that one should evaluate cup products of classes in the image of μ against the fundamental class of $\mathcal{M}_X(P)$. To do this, one first needs to orient $\mathcal{M}_X(P)$. This is accomplished by orienting $H^2_+(X)$ (see [**D5**]). If P is an $SO(3)$ bundle, we fix an integral lift of $w_2(P) \in H^2(X;\mathbf{Z}_2)$ and always identify such a lift with its Poincaré dual $c \in H_2(X;\mathbf{Z})$. The Pontryagin number $p_1(P)$ is congruent to $c^2 \pmod 4$. If c and c' are two integral "lifts" of $w_2(P)$, then the difference in induced orientations is given by $(-1)^{(\frac{c-c'}{2})^2}$. We say that c and c' are equivalent if they are congruent (mod2) and if $(-1)^{(\frac{c-c'}{2})^2} = +1$. The combination of the orientations of X and $H^2_+(X)$ together with an equivalence class c of lifts of $w_2(P)$ is called a "homology orientation" of X. (In case P is an $SU(2)$ bundle, one chooses $c = 0$.) For a Kähler surface X with Kähler class K_X, there is a natural orientation induced from the Kähler structure and a choice of a lift c gives an orientation which differs from this one by $(-1)^{\frac{1}{2}(c^2 + c\cdot K_X)}$ [**D5**].

The moduli space $\mathcal{M}_X(P)$ is, in general, noncompact and needs to be compactified before a fundamental class can be defined. The Uhlenbeck compactification $\overline{\mathcal{M}}_X(P)$ is well-suited to this. However, this compactification is a stratified space and is not usually a manifold. Thus, to define a fundamental class one needs to insure that the singular set has codimension at least 2. This turns out to be the case when either $w_2(P) \neq 0$ or when $w_2(P) = 0$, $d > \frac{3}{4}(1 + b_X^+)$. In practice, one is able to get around this latter restriction by blowing up X and considering bundles over $X \# \overline{\mathbf{CP}}^2$ which are nontrivial when restricted to the exceptional divisor [**MM**].

In [**FM2**] it is shown that for $\alpha \in H_2(X;\mathbf{Z})$ the classes $\mu(\alpha) \in H^2(\mathcal{M}_X(P))$ extend over $\overline{\mathcal{M}}_X(P)$. When b_X^+ is odd, $\dim \mathcal{M}_X(P)$ is even, say equal to $2d$. In fact, a class $c \in H_2(X;\mathbf{Z})$ and a nonnegative integer $d \equiv -c^2 + \frac{1}{2}(1+b^+)$ determine an $SO(3)$ bundle $P_{c,d}$ over X with $w_2(P_{c,d}) \equiv c \pmod 2$ and formal dimension $\dim \mathcal{M}_X(P_{c,d}) = 2d$. For $\bar\alpha = (\alpha_1,\dots,\alpha_d) \in H_2(X;\mathbf{Z})^d$, write $\mu(\bar\alpha) = \mu(\alpha_1)\cup\cdots\cup\mu(\alpha_d)$. Then one has

$$\langle \mu(\bar\alpha), [\overline{\mathcal{M}}_X(P_{c,d})]\rangle = \int_{\overline{\mathcal{M}}_X(P_{c,d})} \mu(\bar\alpha)$$

when $\mu(\bar\alpha)$ is viewed as a $2d$-form.

If $[1] \in H_0(X;\mathbf{Z})$ is the generator, then $\nu = \mu([1]) = -\frac{1}{4}p_1(\beta) \in H^4(\mathcal{M}_X(P))$ where β is the basepoint fibration $\tilde{\mathcal{M}}_X(P) \to \mathcal{M}_X(P)$ with $\tilde{\mathcal{M}}_X(P)$ the manifold of anti-self-dual connections on P modulo based gauge transformations, i.e. those that are the identity on the fiber over a fixed basepoint. The class ν extends over the Uhlenbeck compactification $\overline{\mathcal{M}}_X(P)$ if $w_2(P) \neq 0$, and in case P is an $SU(2)$ bundle, the class will extend under certain dimension restrictions. Once again, these restrictions can be done away with via the tricks mentioned above [**MM**].

Consider the graded algebra

$$\mathbf{A}(X) = \mathrm{Sym}_*(H_0(X) \oplus H_2(X))$$

where $H_i(X)$ has degree $\frac{1}{2}(4-i)$. The Donaldson invariant $D_c = D_{X,c}$ will be an element of the dual algebra $\mathbf{A}^*(X)$, i.e. a linear function

$$D_c : \mathbf{A}(X) \to \mathbf{R}.$$

This is a homology orientation-preserving diffeomorphism invariant for manifolds X satisfying $b_X^+ \geq 3$. Throughout these lectures we assume $b_X^+ \geq 3$ and odd.

Let $x \in H_0(X)$ be the generator [1] corresponding to the orientation. We reserve the use of $1 \in \mathbf{A}(X)$ to denote the unit in degree 0. In case $a + 2b = d > \frac{3}{4}(1+b_X^+)$ and $\alpha \in H_2(X)$,

$$D_c(\alpha^a x^b) = \langle \mu(\alpha)^a \nu^b, [\overline{\mathcal{M}}_X(P_{c,d})]\rangle\,.$$

It will be convenient to extend μ over $\mathbf{A}(X)$, and write for $z \in \mathbf{A}(X)$ of degree d, $D_c(z) = \langle \mu(z), [\overline{\mathcal{M}}_X(P_{c,d})]\rangle$. Since such moduli spaces $\mathcal{M}_X(P_{c,d})$ exist only for $d \equiv -c^2 + \frac{1}{2}(1+b_X^+) \pmod 4$, the Donaldson invariant D_c is defined only on elements of $\mathbf{A}(X)$ whose total degree is congruent to $-c^2 + \frac{1}{2}(1+b_X^+) \pmod 4$. By definition, D_c is 0 on all elements of other degrees. We say that

$$\deg D_c \equiv -c^2 + \frac{1}{2}(1+b_X^+) \pmod 4.$$

When P is an $SU(2)$ bundle one simply writes D or D_X.

A pivotal idea due to Kronheimer and Mrowka [**KM1**] is to consider the invariant

$$\hat{D}_{X,c} : \mathrm{Sym}_*(H_2(X)) \to \mathbf{R}$$

defined by $\hat{D}_{X,c}(u) = D_{X,c}((1+\frac{x}{2})u)$. Whereas $D_{X,c}$ can be nonzero only in degrees congruent to $-c^2 + \frac{1}{2}(1+b^+) \pmod 4$, $\hat{D}_{X,c}$ can be nonzero in degrees congruent to $-c^2 + \frac{1}{2}(1 + b^+)$ (mod 2). The *Donaldson series* $\mathbf{D}_c = \mathbf{D}_{X,c}$ is defined by

$$\mathbf{D}_{X,c}(\alpha) = \hat{D}_{X,c}(\exp(\alpha)) = \sum_{d=0}^{\infty} \frac{\hat{D}_{X,c}(\alpha^d)}{d!}$$

for all $\alpha \in H_2(X)$. This is a formal power series on $H_2(X)$.

The first general properties of the Donaldson invariant were given by Donaldson **[D6]**. These include a vanishing theorem ($D_{X\#Y} \equiv 0$ if both $b^+(X) > 0$ and $b^+(Y) > 0$), and a non-vanishing theorem ($D_X(h^r) > 0$ $(r >> 0)$ if X is an algebraic surface and h is an ample divisor).

We can use these theorems, together with Donaldson's first theorem (Theorem A) to determine, in certain instances, if a 4-manifold is irreducible.

Exercise 7. If X is spin and $D_X \neq 0$, then X is irreducible.

Exercise 8. If X is not spin with $D_X \neq 0$ and $X = X_1 \# X_2$, then one of the X_i has a negative definite intersection form.

Problem 7. Determine if a minimal algebraic surface is irreducible. More generally, when $X = X_1 \# X_2$, determine if one of X_1 or X_2 a connected sum of $\overline{\mathbf{CP}}^2$'s. [3]

The Donaldson invariant D_X is a refinement of the intersection form Q_X, and it too has its incumbent existence and uniqueness questions.

Problem 8. (Partially solved (cf. Theorem 1.3.1)) Which elements $D \in \mathbf{A}^*(X)$ occur as the Donaldson invariant for some smooth 4-manifold X; which formal power series on $H_2(X)$ occur as the Donaldson series of a smooth 4-manifold?

Problem 9. Are there examples of non diffeomorphic simply-connected smooth 4-manifolds X and Y with $D_X = D_Y$. Potential examples are the Horikowa surfaces which are discussed in the last lecture.

Problem 10. Determine canonical procedures for producing all simply-connected smooth 4-manifolds so that the resulting Donaldson invariants can be computed.

From the outset it was clear that the Donaldson polynomial invariants defined for a smooth 4-manifolds X, one for each $SU(2)$-bundle over X, were powerful tools to distinguish homeomorphic but non diffeomorphic smooth 4-manifolds. It was also clear that they were very difficult to compute. Initially, pages of detailed algebraic geometry ([**B**],[**Fr**],[**FM2**],[**MO**]), or analysis and topology ([**GM**],[**MM2**],[**SS**]) led to the computation of just the first one or two non-trivial coefficient of these invariants for special (elliptic) surfaces. This, however, was sufficient to show that the deformation type of elliptic surfaces coincides with their diffeomorphism classification ([**B**],[**MM2**], [**MO**],[**Fr**],[**FS10**]). A (recent) major breakthrough was the discovery of Peter Kronheimer and Tomas Mrowka (Theorem 1.3.1) that there are universal relations that relate these invariants defined on different moduli spaces. Then by considering the generating function determined by these invariants, i.e. the Donaldson series, these relations translated to, under a technical assumption called simple-type (defined below), a differential equation satisfied by the Donaldson series. This gave structure to the seemingly disparate Donaldson invariants. At bottom Kronheimer and Mrowka showed that for each smooth 4-manifold with

---

[3] Added in proof. Every minimal symplectic 4-manifold is irreducible: cf. Taubes, The Seiberg-Witten and Gromov invariants, Math. Res. Letters (2) 1996, 221-238.

vanishing first homology there are a finite number of two-dimensional cohomology classes and rational numbers which completely determine its Donaldson series.

Many of these results can be also understood by studying how D_X behaves when evaluated on certain powers of homology classes $x \in H_2(X;\mathbf{Z})$ which are represented by spheres.

Studying embedded spheres appears to be overly restrictive; however:

Exercise 9. If X is simply-connected and $x \in H_2(X;\mathbf{Z})$ arbitrary, then x is represented by an immersed sphere with ordinary double points.

Exercise 10. If $x \in H_2(X;\mathbf{Z})$ is represented by an immersed sphere with p_x positive double points and n_x negative double points, then $x + 2e_1 + \cdots + 2e_{p_x}$ is represented by an embedded sphere in $X \#_{p_x+n_x} \overline{\mathbf{CP}}^2$. Here, the e_j are the exceptional divisors of the $\overline{\mathbf{CP}}^2$, i.e. generators of $H_2(\overline{\mathbf{CP}}^2;\mathbf{Z})$ with $Q(e_j,e_j) = -1$.

Exercise 11. If $D_X \neq 0$ and $x \in H_2(X;\mathbf{Z})$ is represented by an embedded sphere then $Q(x,x) \leq 0$. A more advanced exercise is to show that if $x \neq 0$, then $Q(x,x) < 0$.

Exercise 12. Construct a 4-manifold X with $b_1(X) = 0$ for which a basis of $H_2(X;\mathbf{Q})$ is not represented by immersed 2-spheres.

For this approach our first task will be to relate D_X with $D_{X\#\overline{\mathbf{CP}}^2}$; this we do in Lecture 2 (cf. [**FS8**]). This together with an inductive construction of $E(n)$ from $E(n-1)\#\overline{\mathbf{CP}}^2$ will allow us to inductively compute $D_{E(n)}$ and get the ball rolling. In particular, we show $\mathbf{D}_{E(n)} = \exp(Q/2)\sinh^{n-2}(F)$ as formal power series, where F is the cohomology class of the fiber.

Many of the gauge-theoretic results we need are proved using "neck-stretching" arguments which involve splitting a 4-manifold along a 3-manifold and analyzing how the moduli spaces of anti-self-dual connections over the 4-manifold decompose into pieces. The complexity of these arguments is directly proportional to the complexity of the character variety of the 3-manifold along which the 4-manifold is split. In all our arguments we split along the most elementary 3-manifolds, the lens spaces $L(p,q)$, with the most elementary nontrivial character variety, a finite collection of points. This accounts for the relative simplicity of the arguments.

As we stated above, the relations derived for embedded spheres (cf. Theorem 2.3.1) suffice to answer Problem 8, at least in the case X has simple-type. We refer the audience to the original research article for exactly how this is accomplished [**FS9**]. A simply connected 4-manifold X is said to have *simple type* if the relation $D_{X,c}(x^2\, z) = 4\, D_{X,c}(z)$ is satisfied by its Donaldson invariant for all $z \in \mathbf{A}(X)$ and for all $c \in H_2(X;\mathbf{Z})$. As we already mentioned, this important definition is due to Kronheimer and Mrowka [**KM1**] and was observed to hold for many 4-manifolds. In terms of $\hat{D}_{X,c}$, the simple type condition is that $\hat{D}_{X,c}(zx) = 2\hat{D}_{X,c}(z)$ for all $z \in \mathbf{A}(X)$ and for all $c \in H_2(X;\mathbf{Z})$. The assumption of simple type assures that for each c, the complete Donaldson invariant $D_{X,c}$ is determined by the Donaldson series $\mathbf{D}_{X,c}$. It is still an open question whether all 4-manifolds are of simple type.

The main structure theorem for the Donaldson series is:

Theorem 1.3.1 (Kronheimer and Mrowka [**KM1, KM2, FS9**]). *Let X be a simply connected 4-manifold of simple type. Then, there exist finitely many homology* *classes* $\kappa_1, \ldots, \kappa_p \in H_2(X, \mathbf{Z})$ *and nonzero rational numbers* $a_1, \ldots, a_p$ *such that*

$$\mathbf{D}_X = \exp(Q/2) \sum_{s=1}^{p} a_s e^{\kappa_s}$$

as analytic functions on $H_2(X)$. *Each of the 'basic classes'* κ_s *is characteristic, i.e.* $\kappa_s \cdot x \equiv x \cdot x \pmod 2$ *for all* $x \in H_2(X; \mathbf{Z})$.

Further, suppose $c \in H_2(X; \mathbf{Z})$. *Then*

$$\mathbf{D}_{X,c} = \exp(Q/2) \sum_{s=1}^{p} (-1)^{\frac{c^2 + \kappa_s \cdot c}{2}} a_s e^{\kappa_s}$$

Here the homology class κ_s acts on an arbitrary homology class by intersection, i.e. $\kappa_s(u) = \kappa_s \cdot u$.

Theorem 1.3.2 (Kronheimer and Mrowka [**KM1, KM2**]). *If* $u \in H_2(X; \mathbf{Z})$ *is represented by an embedded surface of genus* g *with self-intersection* $u^2 \geq 0$, *then for each* s

$$2g - 2 \geq u^2 + |\kappa_s \cdot u|.$$

Theorem 1.3.3 ([**FS9**]). *Let* X *be a simply connected 4-manifold of simple type and let* $\{\kappa_s\}$ *be the set of basic classes as above. If* $u \in H_2(X; \mathbf{Z})$ *is represented by an immersed 2-sphere with* $p \geq 1$ *positive double points, then for each* s

$$2p - 2 \geq u^2 + |\kappa_s \cdot u|. \tag{1}$$

Theorem 1.3.4 ([**FS9**]). *Let* X *be a simply connected 4-manifold of simple type with basic classes* $\{\kappa_s\}$ *as above. If the nontrivial class* $u \in H_2(X; \mathbf{Z})$ *is represented by an immersed 2-sphere with no positive double points, then let*

$$\{\kappa_s |\, s = 1, \ldots, 2m\}$$

be the collection of basic classes which violate the inequality (1). Then $\kappa_s \cdot u = \pm u^2$ *for each such* κ_s. *Order these classes so that* $\kappa_s \cdot u = -u^2\ (> 0)$ *for* $s = 1, \ldots, m$. *Then*

$$\sum_{s=1}^{m} a_s e^{\kappa_s + u} - (-1)^{\frac{1+b_X^+}{2}} \sum_{s=1}^{m} a_s e^{-\kappa_s - u} = 0.$$

Problem 11. Determine whether the basic classes κ_s satisfy $\kappa_s^2 = 3\sigma(X) + 2e(X)$. [4]

Problem 12. Determine whether, up to multiples, the basic classes represent the first Chern class of some symplectic structure on X. [5]

Problem 13. Does X have big diffeomorphism group with respect to the basic classes κ_s?

Problem 14. Determine whether every simply-connected smooth 4-manifold with $b^+ > 1$ has simple-type.

Known examples suggest:

---

[4] Added in proof. True for symplectic manifolds (Taubes, The Seiberg-Witten and Gromov invariant, Math. Res. Letters 2 (1995), 221-238).

[5] Added in proof. False in general, cf. footnote to Problem 5. For symplectic manifolds, there are some restrictions (cf. Taubes).

Problem 15. Suppose that X has simple-type with two basic classes $\pm\kappa$. Then determine whether

$$\mathbf{D}_X = 2^{3\sigma+2e-\frac{b^+-3}{2}} \exp(Q/2)\,(e^{\kappa_s} \pm e^{-\kappa_s})$$

In the course of these lectures we will determine certain criteria for when X has simple-type, the most useful being variants of Theorem 3.2.1.

As noted above, the largest known simply-connected irreducible smooth 4-manifolds with a given b^+ are the elliptic surfaces $E(n)$. If some of our problems above have an affirmative answer (e.g. $2e(X) - 3|\sigma(X)| \geq 0$) they indeed are the largest, so that we need to find a method starting with the $E(n)$ and their known Donaldson invariants to make them "smaller" in such a fashion that we can keep track of the Donaldson invariants. The last two lectures are devoted to this. Note that by blowing down an exceptional curve, b^+ remains constant, but c_1^2 increases. Of course irreducible 4-manifolds cannot be blown down. What we need to do is concoct a rational version of blowing down a smooth 4-manifold. We do this as follows.

Let C_p denote the simply-connected smooth 4-manifold obtained by plumbing the $(p-1)$ disk bundles over the 2-sphere according to the linear diagram

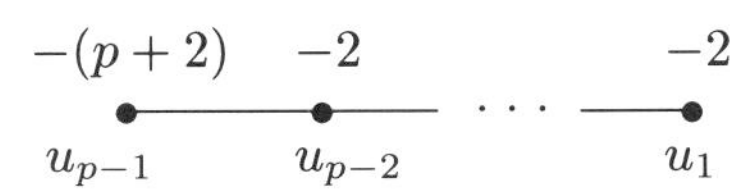

Here, each node denotes a disk bundle over S^2 with Euler class indicated by the label; an interval indicates that the endpoint disk bundles are plumbed, i.e identified fiber to base over the upper hemisphere of each S^2. Label the homology classes represented by the spheres in C_p by $u_1, \dots, u_{p-1}$ so that $u_{p-1}^2 = -(p+2)$ and, for $j = 1, \dots, p-2$, $u_j^2 = -2$. Further, orient the spheres so that $u_j \cdot u_{j+1} = +1$. Then C_p is a 4-manifold with negative definite intersection form and with boundary the lens space $L(p^2, p-1)$.

Exercise 13. $L(p^2, p-1)$ bounds a rational ball B_p with $\pi_1(B_p)) = \mathbf{Z}/p\mathbf{Z}$ and with the inclusion induced homomorphism $\pi_1(\partial B_p) \to \pi_1(B_p)$ the obvious surjection.

(Hint: There are several ways to construct B_p; here is an outline of one. Let $\mathbf{F}_{p-1}$, $p \geq 2$, be the simply connected ruled surface whose negative section s_- has square $-(p-1)$. Let s_+ be a positive section (with square $(p-1)$) and f a fiber. Then the homology classes $s_+ + f$ and s_- are represented by embedded 2-spheres which intersect each other once and have intersection matrix

$$\begin{pmatrix} p+1 & 1 \\ 1 & -(p-1) \end{pmatrix}$$

It follows that the regular neighborhood of this pair of 2-spheres has boundary $L(p^2, p-1)$. Its complement in $\mathbf{F}_{p-1}$ is the rational ball B_p.)

The above construction is amenable to showing:

Problem 16. If the configuration of spheres C_p are symplectically embedded in a symplectic 4-manifold X, then the rational blowdown X_p is also symplectic (cf. [**G3**]). [6]

[6] Added in proof. This has been proven by John Etynve in his 1997 U. Texas thesis.

Here is another hint to show how to construct B_p in Exercise 13. The second construction begins with the configuration of $(p-1)$ 2-spheres

in $(p-1)\mathbf{CP}^2$ where the spheres (from left to right) represent

$$2h_1 - h_2 + \cdots + h_{p-1},\ h_1 + h_2,\ h_2 + h_3, \ldots, h_{p-2} + h_{p-1}$$

where h_i is the hyperplane class in the ith copy of $\mathbf{CP}^2$. The boundary of the regular neighborhood of the configuration is $L(p^2, p-1)$, and the classes of the configuration span $H_2(\mathbf{CP}^2; \mathbf{Q})$. The complement is the rational ball B_p.

Suppose that C_p embeds in a closed smooth 4-manifold X. Then let X_p be the smooth 4-manifold obtained by removing the interior of C_p and replacing it with B_p. We call this procedure a **rational blowdown** and say that X_p is obtained by **rationally blowing down** X. Note that $b^+(X) = b^+(X_p)$ so that rationally blowing down increases the signature while keeping b^+ fixed.

The principal gauge-theoretic result here is:

Theorem 1.3.5 ([**FS10**]). *Suppose that X has simple type and*

$$\mathbf{D}_X = \exp(Q/2) \sum_{s=1}^{n} a_s e^{\kappa_s}.$$

Let $C_p \subset X$ and let X_p be its rational blowdown. Then

$$\mathbf{D}_{X_p} = \exp(\bar{Q}/2) \sum_{s=1}^{n} b_s e^{\bar{\kappa}_s}$$

where the b_s depend only on the intersection numbers $u_i \cdot \kappa_s$, $i = 1, \ldots, p-1$, and $b_s = 0$ unless $\partial \kappa'_s \in p\mathbf{Z}_{p^2}$. The basic classes $\bar{\kappa}_s \in H_2(X_p; \mathbf{Z})$ are the unique extensions of the κ'_s. □

In short, the Donaldson invariants for X_p are completely determined by those of X; i.e there are rational blowdown formulas.

Problem 17. Determine rational blowup formulas.

Many such configurations embed in the $E(n)$ so that we can construct several manifolds in the regions graphed in Figure 1 and compute all their Donaldson invariants. It turns out that another surgical procedure called a topological logarithmic transformation of order p can be viewed as a rational blowdown of a naturally occuring $C(p)$ in $X \#_{p-1} \overline{\mathbf{CP}}^2$ so that the blowup formula and the rational blowdown formula computes the resulting Donaldson invariants (cf. [**FS10**]). In particular we can compute all the Donaldson invariants of the elliptic surfaces $E(n; m_1, \ldots m_r)$ with multiple fibers of order m_j. In particular

$$\mathbf{D}_{E(n;m_1,\ldots m_r)} = \exp\left(\frac{Q}{2}\right) \frac{(\sinh F)^{n+r-2}}{\prod_i \sinh(F/m_i)}.$$

As an amusing corollary we compute the Donaldson series of the Gompf-Mrowka examples $K = K(p_1, q_1; p_2, q_2; p_3, q_3)$ [**GM**] and find that

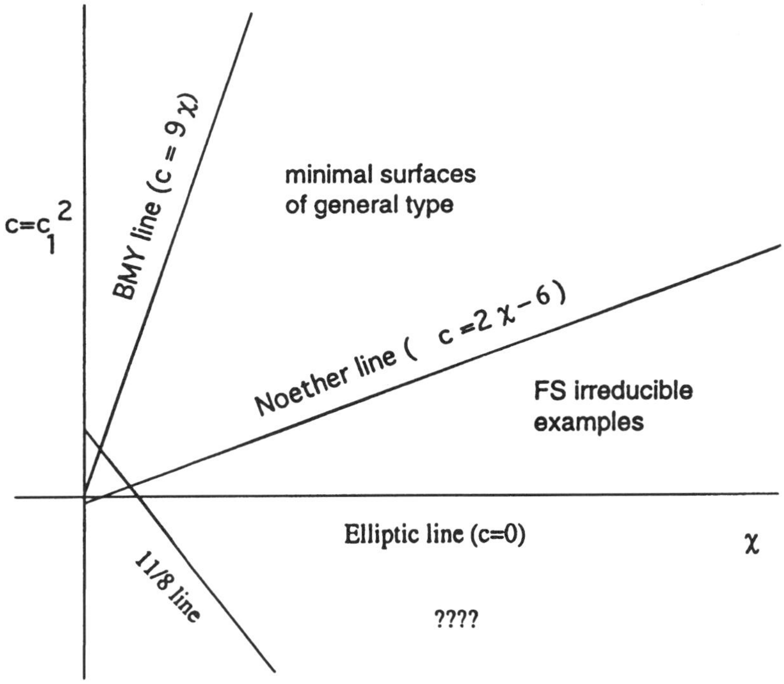

Figure 1. Geography of 4-manifolds

$$\mathbf{D}_K = \exp\left(\frac{Q}{2}\right) \prod_{j=1}^{3} \frac{(\sinh F_j)^2}{\sinh(F_j/p_j)\sinh(F_j/q_j)} .$$

Now lets get to work.

LECTURE 2
−2 Spheres and the Blowup Formula

2.1. Ruberman's theorem

We begin by studying the behavior of the Donaldson invariant of a 4-manifold with a homology class τ represented by an embedded 2-sphere S of self-intersection $\tau \cdot \tau = -2$. Let $\langle \tau \rangle^\perp$ denote $\{\alpha \in H_2(X) | \tau \cdot \alpha = 0\}$ and let

$$\mathbf{A}(\tau^\perp) = \mathbf{A}_X(\tau^\perp) = \mathrm{Sym}_*(H_0(X) \oplus \langle \tau \rangle^\perp).$$

Theorem 2.1.1. (Ruberman [**R**]) *Suppose that $\tau \in H_2(X; \mathbf{Z})$ with $\tau \cdot \tau = -2$ is represented by an embedded sphere S. Then for $z \in \mathbf{A}(\tau^\perp)$, we have $D(\tau^2 z) = 2\, D_\tau(z)$.*

Proof. Write $X = X_0 \cup N$ where N is a tubular neighborhood of S, and note that ∂N is the lens space $L(2,-1) = \mathbb{R}P^3$. Since $b^+(X_0) > 0$, generically there are no reducible anti-self-dual (ASD) connections on X_0. However, since $b^+(N) = 0$ there are indeed nontrivial reducible ASD connections in complex line bundles λ^m, $m \in \mathbb{Z}$, with $c_1(\lambda^m) = m$ represented by a harmonic 2-form and with $\lambda^m|_{\partial N}$ the flat line bundle with holonomy -1 on the meridian curve, i.e. $\langle c_1(\lambda), \tau \rangle = -1$. In particular λ is determined by the non-trivial representation ζ which generates the character variety of $SU(2)$ representations of $\pi_1(\partial N) = \mathbf{Z}_2$ mod conjugacy. (Of course, ζ^{2m} is trivial, and $\zeta^{2m+1} = \zeta$.) Let $\mathbf{C} \cong \lambda^0$ be the trivial bundle. Note that for reducible connections on N, $c_1 \in H^2(N; \mathbb{Z}) \cong H_2(N; \partial N; \mathbb{Z}) \cong \mathbb{Z}$ and that $c_1(\lambda) = \gamma$ is a generator. Then $-4p_1(\lambda) = c_2(\lambda \oplus \bar{\lambda}) = -c_1^2(\lambda) = \frac{1}{2}$.

The structure of the moduli spaces $\mathcal{M}_N(\lambda^m \oplus \bar{\lambda}^m)$ and $\mathcal{M}_{X_0}$ have been studied in [**MMR**] and [**T1**]. In our setting (the boundaries are lens spaces) an anti-self-dual connection has a well-defined limiting flat connection on the restriction of the bundle to $\partial N = -\partial C_0$ and, for a generic metric on X_0, the moduli space of anti-self-dual connections $\mathcal{M}_{c_{X_0}}[\zeta^m]$ on X_0 with energy c_{X_0} and asymptotic value ζ^m contains no reducible connections and is a smooth manifold of dimension

$$\dim \mathcal{M}_{c_{X_0}}[\zeta^m] = 8c_{X_0} - \frac{3}{2}(e(X_0) + \sigma(X_0)) - \frac{h_{\zeta^m}}{2} - \frac{\rho_{\zeta^m}}{2} = 8c_{X_0} - 3(1 + b^+(X)).$$

Further, $\mathcal{M}_N(\lambda^m \oplus \bar{\lambda}^m)$ is a manifold away from the reducible connection with dimension

$$\dim \mathcal{M}_{c_N}(N) = 8c_N - \frac{3}{2}(e(N) + \sigma(N)) - \frac{h_{\zeta^m}}{2} + \frac{\rho_{\zeta^m}}{2} = 8c_N - 3 = 4m^2 - 3.$$

Here, h_{ζ^m} is the dimension of the subgroup of the gauge group which stabilizes the flat connection ζ^m (=0 in our case) and ρ_{ζ^m} is its rho-invariant (=0 in our case).

We now invoke what is called a "neck-stretching" argument. That is, consider a sequence of metrics g_n on X containing increasingly long tubes $[-T, T] \times \partial N$ and determine what happens to a sequence of anti-self-dual connections in $\mathcal{M}(X, g_i) \cap V_1 \cap \ldots V_{d-2} \cap V_{\tau,1} \cap V_{\tau,2}$, where the V's are divisors associated to the $\mu(z)$ and $\mu(\tau)$. If $A_i \in \mathcal{M}(X, g_i) \cap V_1 \cap \ldots V_{d-2} \cap V_{\tau,1} \cap V_{\tau,2}$, then stretching the neck we get that in the Uhlenbeck limit $A_i \mapsto A_{X_0} \amalg A_N + r$ bubbles on $X_0 + s$ bubbles on $N + t$ tubes $\partial N \times \mathbf{R}$, where $A_{X_0} \in \mathcal{M}_{X_0}$ $A_N \in \mathcal{M}_N$. Further $\dim \mathcal{M}_{X_0} + \dim \mathcal{M}_N + 8r + 8s + 3 \leq 2d$.

The next step is a sequence of dimension counting arguments to determine r, s, and the irreducibility of A_{X_0} and A_N.

If A_{X_0} and A_N are both irreducible, then $\dim \mathcal{M}_{X_0} \geq 2(d - 2 - 2r)$ and $\dim \mathcal{M}_N \geq 2(2 - 2s)$ so that $(2d - 4 - 4r) + (4 - 4s) + 8r + 8s + 3 \leq 2d$ which is impossible.

If A_{X_0} is flat and A_N is irreducible, then $2r \geq d - 2$ and $\dim \mathcal{M}_{X_0} = 0 - 3(1 + b^+(Q))$ so that $-3(1 + b^+(X_0)) + (4 - 4s) + (4d - 8) + 8s + 3 \leq 2d$; i.e. $2d - 3(1 + b^+(X)) - 1 + 4s \leq 0$. But to define the Donaldson invariants we need that $k > \frac{3}{4}(1 + b^+(X)) + \frac{1}{2}$, so that $2d - 3(1 + b^+(X)) > 4$. So we cannot have A_{X_0} flat and A_N irreducible.

If A_{X_0} is flat and A_N is flat (hence trivial), then $s \geq 1$ and $\dim \mathcal{M}_N = -3$ so that $-3(1 + b^+(X)) + (-3) + (4d - 8) + 8 + 3 \leq 2d$; i.e. $2d - 3(1 + b^+(X)) \leq 0$ which again contradicts $k > \frac{3}{4}(1 + b^+(X)) + \frac{1}{2}$.

If A_{X_0} is flat and A_N is reducible and not flat, then A_N is the reducible ASD connection in $\lambda^m \oplus \overline{\lambda}^m$ with $c_2(\lambda^m \oplus \overline{\lambda}^m) = \frac{m^2}{2}$. Then $\dim \mathcal{M}_N = 4m^2 - 3 \geq 1$. Then, again, $-3(1 + b^+(X)) + 1 + (4d - 8) + 8s + 3 \leq 2d$; i.e. $2d - 3(1 + b^+(X)) - 4 + 8s \leq 0$ contradicts $k > \frac{3}{4}(1 + b^+(X)) + \frac{1}{2}$.

If A_{X_0} is irreducible and A_N is flat (hence trivial), then $(2d - 4 - 4r) + (-3) + 8r + 8 \leq 2d$; i.e. $4r + 1 \leq 0$, a contradiction.

There is but one situation left: **A_{X_0} is irreducible and A_N is reducible and not flat**. This will be the common theme for the proofs of many of our arguments; counting arguments will get one into the situation where the big side has an irreducible connection and the negative definite side has a reducible nonflat connection.

Since $\langle\tau\rangle^\perp = H_2(X_0)$, we need to evaluate the Donaldson invariant on two copies of τ and classes in $H_2(X_0)$. After cutting down moduli spaces by intersecting with transverse divisors representing the images under μ of these classes in $H_2(X_0)$ and using the given homology orientation, we may assume without loss of generality that there are no such classes and that we are working with a 4-dimensional moduli space $\mathcal{M}_X$. Let V_1 and V_2 be divisors representing $\mu(\tau)$, coming from general positioned surfaces in N. The Donaldson invariant is the signed intersection number

$$D(\tau^2) = \#(\mathcal{M}_X \cap V_1 \cap V_2).$$

A standard dimension counting argument (cf. [**D6**]) shows that if we choose a metric on X with long enough neck length, $\partial N \times [0,T]$, then all the intersections take place in a neighborhood $\mathcal{U}$ of the grafted moduli space $\mathcal{M}_{X_0}[\zeta]\#\{A_\lambda\}$ where A_λ is the reducible anti-self-dual connection on $\lambda \oplus \bar{\lambda}$, and $\mathcal{M}_{X_0}[\zeta]$ is the 0-dimensional cylindrical end moduli space on X_0 consisting of anti-self-dual connections which decay exponentially to the boundary value ζ. Let m_{X_0} be the signed count of points in $\mathcal{M}_{X_0}[\zeta]$. A neighborhood of A_λ in the moduli space $\mathcal{M}_N(\lambda \oplus \bar{\lambda})$ is diffeomorphic to $(\mathbf{C} \times_{S^1} SO(3))/SO(3) \cong \mathbf{C}/S^1 \cong [0,\infty)$. Here S^1 acts on $SO(3)$ so that $SO(3)/S^1 = S^2$ and on $\mathbf{C}$ with weight -2. Thus the neighborhood $\mathcal{U}$ is

$$(\tilde{\mathcal{M}}_{X_0}[\zeta] \times (\mathbf{C} \times_{S^1} SO(3)))/SO(3)$$

where "$\tilde{\mathcal{M}}_{X_0}[\zeta]$" denotes the based moduli space.

Now $\tilde{V}_1 \cap (\mathbf{C} \times_{S^1} SO(3)) = \{0\} \times_{S^1} SO(3)$, and the intersection of V_1 with all of $\mathcal{M}_X$ is

$$(\mathcal{M}_{X_0}[\zeta] \times (\{0\} \times_{S^1} SO(3)))/SO(3) = \Delta.$$

Fix a point $p \in \mathcal{M}_{X_0}[\zeta]$, let $SO(3) \cdot p$ denote its orbit in $\tilde{\mathcal{M}}_{X_0}[\zeta]$, and let

$$\Delta_p = (SO(3) \cdot p \times (\{0\} \times_{S^1} SO(3)))/SO(3) \cong S^2.$$

Identify Δ_p with a transversal in $\tilde{\Delta}_p$ and compute the intersection number $\tilde{V}_2 \cdot \Delta_p = \iota_p$. Since ι_p is independent of $p \in \mathcal{M}_{X_0}[\zeta]$, we have $D(\tau^2) = \iota_p \cdot m_{X_0}$. The constant ι_p is computed in [**FM2**] as follows. Note that $\Delta_p = \{0\} \times_{S^1} SO(3) \subset \mathbf{C} \times_{S^1} SO(3)$ is a zero-section of the $c_1 = -2$ complex line bundle over S^2 and $\tilde{V}_2$ is another section. Thus $\tilde{V}_2 \cdot \Delta_p = -2$; and so $D(\tau^2) = -2\, m_{X_0}$.

To identify the relative invariant m_{X_0}, view $\mathcal{M}_{X_0}[\zeta]$ as $\mathcal{M}_{X_0,0}[\mathrm{ad}(\zeta)]$, an $SO(3)$ moduli space. Since $\mathrm{ad}(\zeta)$ is the trivial $SO(3)$-representation, we may graft connections in $\mathcal{M}_{X_0,0}[\mathrm{ad}(\zeta)]$ to the trivial $SO(3)$ connection over N, and since $b_N^+ = 0$, there is no obstruction to doing this. We obtain an $SO(3)$ moduli space over X corresponding to an $SO(3)$ bundle over X with w_2 Poincaré dual to τ. (This is the unique nonzero class in $H^2(X;\mathbf{Z}_2)$ which restricts trivially to both N and X_0.) Thus for $z \in \mathbf{A}(\tau^\perp)$, we have $D(\tau^2\, z) = \pm 2\, D_\tau(\, z)$. (Note that since $\tau \cdot \tau = -2$, we have $D_{-\tau} = D_\tau$.)

To determine the sign is an (important) technical matter that we refer the reader to [**FS8**].

For the case of the $SO(3)$ invariants the proof of Theorem 2.1.1 can be easily adapted to show:

Theorem 2.1.2. *Suppose that $\tau \in H_2(X;\mathbf{Z})$ with $\tau \cdot \tau = -2$ is represented by an embedded sphere S. Let $c \in H_2(X;\mathbf{Z})$ satisfy $c \cdot \tau \equiv 0 \pmod 2$. Then for $z \in \mathbf{A}(\tau^\perp)$ we have $D_c(\tau^2\, z) = 2\, D_{c+\tau}(z)$.* □

These same "neck-stretching" and "dimension counting" techniques can be utilized to prove some elementary facts concerning the Donaldson invariants of blowups. These can be found, for example in [**FM2, Ko, L**]. Let X have the Donaldson invariant D, and let $\hat{X} = X\#\overline{\mathbf{CP}}^2$ have the invariant $\hat{D}$.

Exercise 14. Let $e \in H_2(\overline{\mathbf{CP}}^2;\mathbf{Z}) \subset H_2(\hat{X};\mathbf{Z})$ be the exceptional class, and let $c \in H_2(X;\mathbf{Z})$. Then for all $z \in \mathbf{A}(X)$:

1. $\hat{D}_c(e^{2k+1}\, z) = 0$ for all $k \geq 0$.

2. $\hat{D}_c(z) = D_c(z)$.
3. $\hat{D}_c(e^2 z) = 0$.
4. $\hat{D}_c(e^4 z) = -2\, D_c(z)$.
5. $\hat{D}_{c+e}(e^{2k} z) = 0$ for all $k \geq 0$.
6. $\hat{D}_{c+e}(e\, z) = D_c(z)$.
7. $\hat{D}_{c+e}(e^3 z) = -D_c(x\, z)$.

Item (7) doesn't exactly follow as the proof of Theorem 2.1.1 and uses a key idea that is generalized in [**FS9**]. Here's a sketch of the proof.

Consider a neighborhood N of the exceptional curve, and let $X_0 = \hat{X} \setminus N \cong X \setminus B^4$. As in the proof of Ruberman's theorem we lose no generality by assuming that we are evaluating $\hat{D}$ only on e^3. A dimension counting argument shows that if we stretch the neck between X_0 and N to have infinite length by taking a sequence of generic metrics $\{g_n\}$, and if V_i are transverse divisors representing $\mu(e)$, then any sequence of connections

$$A_n \in \mathcal{M}_{\hat{X},c+e}(g_n) \cap V_1 \cap V_2 \cap V_3$$

must converge to the sum of a connection in the 4-manifold $\mathcal{M}_{X_0,c}$ and the unique reducible connection on N corresponding to the line bundle λ over N whose Euler class is Poincaré dual to e. The based moduli space $\tilde{\mathcal{M}}_N(\lambda)$ is the orbit of this reducible connection, a 2-sphere, S^2_e. Let v denote the (positive) generator of the equivariant cohomology $H^*_{SO(3)}(S^2_e) \cong H^*(\mathbf{CP}^\infty)$ in dimension 2. The class $\mu(e)$ lifts to the equivariant class $-\frac{1}{2}\langle c_1(\lambda), e\rangle\, v = \frac{1}{2} v \in H^2_{SO(3)}(S^2_e)$. The connections in S^2_e are asymptotically trivial and this induces an $SO(3)$ equivariant push-forward map

$$\partial_*(N) : H^*_{SO(3)}(S^2_e) \to H^*_{SO(3)}(1) = H^*(BSO(3)).$$

If $u \in H^*_{SO(3)}(1)$ is the generator in dimension 4 then $\partial_*(N)(v^{2k+1}) = 2\, u^k$. So $\partial_*(N)(e^3) = \frac{1}{4} u$.

Since each connection in $\mathcal{M}_{X_0,c}$ is also asymptotically trivial, there is an induced map $\partial^*(X_0) : H^*_{SO(3)}(1) \to H^*_{SO(3)}(\tilde{\mathcal{M}}_{X_0,c})$. It follows from [**T3, AB**] that $\hat{D}_{c+e}(e^3)$ is obtained by evaluating

$$\begin{aligned} \langle \partial^*(X_0)\, \partial_*(N)(e^3), [\tilde{\mathcal{M}}_{X_0,c}]\rangle &= \frac{1}{4}\langle \partial^*(X_0)(u), [\tilde{\mathcal{M}}_{X_0,c}]\rangle \\ &= \frac{1}{4}\langle \pi_*\partial^*(X_0)(u), [\mathcal{M}_{X_0,c}]\rangle \end{aligned}$$

where basepoint fibration β over X_0 is

$$\pi : \tilde{\mathcal{M}}_{X_0,c} \to \mathcal{M}_{X_0,c}$$

the last equality because the $SO(3)$ action on $\tilde{\mathcal{M}}_{X_0,c}$ is free. But Austin and Braam [**AB**], for example, show that $\pi_*\partial^*(X_0)(u) = p_1(\beta)$. Since $\nu = -\frac{1}{4}p_1(\beta)$, we get $\langle \nu, [\mathcal{M}_{X_0,c}]\rangle = -D_c(x)$.

2.2. The recursion scheme

Surprisingly, we are now in a position to show that there is a recursion scheme for determining the relation between D and $\hat{D}$, as well as a relation between D_{c+e}

and $\hat{D}_{c+e}$. To see this we establish some notation. Let $\hat{X} = X\#\overline{\mathbf{CP}}^2$ and let $e \in H_2(X)$ denote the homology class of the exceptional divisor. Since $b^+_X = b^+_{\hat{X}}$, the corresponding Donaldson invariants $D = D_X$ and $\hat{D} = D_{\hat{X}}$ have their (possible) nonzero values in the same degrees $\pmod 4$. We recursionly show that there are polynomials $B_k(x)$ satisfying

$$\hat{D}(e^k\, z) = D(B_k(x)\, z)$$

and polynomials $S_k(x)$ such that

$$\hat{D}_e(e^k) = D(S_k(x)).$$

Exercise 14 can be restated as

Exercise 15. Compute B_0, B_2, S_1, S_3, B_{2i+1}, and S_{2i}.

Now blowup X twice. The class $e_1 - e_2$ is represented by a sphere with self-intersection -2; so use Theorem 2.1.1 and Theorem 2.1.2 to show

Exercise 16. $\hat{D}_{c+e_1+e_2}((e_1+e_2)^{k+2}(e_1-e_2)^2) = 2\hat{D}_c((e_1+e_2)^{k+2})$.

Exercise 17. $\hat{D}_c((e_1+e_2)^{k+2}(e_1-e_2)^2) = 2\hat{D}_{c+e_1+e_2}((e_1+e_2)^{k+2})$.

Exercise 18. S_{k+3} is determined by $S_1, \ldots, S_{k+1}, B_0, \ldots B_{k+2}$ for all even $k \geq 2$

Exercise 19. B_{k+4} is determined by $B_0, \ldots, B_{k+2}, S_1, \ldots S_{k+1}$ for all even $k \geq 0$. (Hint: Expand both sides of Exercises 16 and 17)

Using Exercise 15 to start the induction, there is now a recursion scheme which determines $B_k(x)$ and $S_k(x)$. Here's a Mathematica exercise.

Exercise 20. Show $B_{12} = -512\,x^4 - 960\,x^2 - 408$ and

$$\begin{aligned} B_{30}(x) = 134,217,728\,x^{13} &+ 4,630,511,616\,x^{11} + 68,167,925,760\,x^9 \\ &- 34,608,135,536,640\,x^7 - 39,641,047,695,360\,x^5 \\ &- 9,886,101,110,784\,x^3 + 543,185,367,552\,x \end{aligned}$$

To actually find a closed form for $B_k(x)$ and $S_k(x)$ requires a bit more work and for this we use a result for embedded 2-spheres of self-intersection -3. Again, it's proof similar to that of Theorem 2.1.1.

Theorem 2.2.1. *Suppose that $\tau \in H_2(X;\mathbf{Z})$ is represented by an embedded 2-sphere S with self-intersection -3. Let $\omega \in H_2(X;\mathbf{Z})$ satisfy $\omega \cdot \tau \equiv 0 \pmod 2$. Then for all $z \in \mathbf{A}(\tau^\perp)$ we have*

$$D_\omega(\tau\, z) = -D_{\omega+\tau}(\,z).$$

To conclude we use the relations given in Theorem 2.1.1, Theorem 2.1.2 and Theorem 2.2.1 to obtain a relation key to solving the recursion scheme given in above. This relation was first proved by Wojciech Wieczorek using different methods. His proof will appear in his thesis [**Wk**].

Corollary 2.2.2. *Suppose that* $\tau \in H_2(X;\mathbf{Z})$ *is represented by an embedded 2-sphere with self-intersection* -2, *and let* $c \in H_2(X;\mathbf{Z})$ *with* $c\cdot\tau \equiv 0 \pmod 2$. *Then for all* $z \in \mathbf{A}(\tau^\perp)$

$$D_c(\tau^4\, z) = -4\, D_c(\tau^2\, x\, z) - 4\, D_c(\, z).$$

Proof. In $\hat{X} = X\#\overline{\mathbf{CP}}^2$ the class $\tau + e$ is represented by a 2-sphere of self-intersection -3, and $(\tau - 2e)\cdot(\tau+e) = 0$. From Lemma 14 we get

$$\hat{D}_c((\tau-2e)^3\,(\tau+e)\,z) = D_c(\tau^4\,z) - 8\,\hat{D}_c(e^4\,z) = D_c(\tau^4\,z) + 16\,D_c(\,z).$$

On the other hand, by Theorems 2.1.2 and 2.2.1 and by Lemma 14,

$$\begin{aligned} \hat{D}_c((\tau-2e)^3\,(\tau+e)\,z) &= -\hat{D}_{c+\tau+e}((\tau-2e)^3\,z) \\ &= 6\,\hat{D}_{c+\tau+e}(\tau^2\,e\,z) + 8\,\hat{D}_{c+\tau+e}(e^3\,z) \\ &= 6\,D_{c+\tau}(\tau^2\,z) - 8\,D_{c+\tau}(x\,z) = 12\,D_c(\,z) - 4\,D_c(\tau^2\,x\,z) \end{aligned}$$

and the result follows.

2.3. Universal relations

This sequence of arguments, starting with Theorem 2.1.1 and ending with Theorem 2.2.2 is a prototype for determining relations for embedded spheres with arbitrary self-intersection. By using neck-stretching arguments (which are complicated by the fact that the moduli spaces are no longer compact) one proves relations such as:

Theorem 2.3.1. *(*[**FS9**]*) Let* X *be an oriented simply connected 4-manifold. If* σ *is a homology class with self-intersection* $\sigma^2 \le -2$ *which is represented by an embedded 2-sphere* S, *then there are constants* $A_{j,b,k}$ *and* $B_{j,b,k}$ *depending only on* σ^2, *such that for* $z \in \mathbf{A}(\sigma^\perp)$

$$D(\sigma^{2k-1}z) = B_{0,0,k}D_\sigma(z) + \sum_{j=1}^{k-1}\sum_{b=0}^{[\frac{k-j}{2}]} B_{j,b,k}D(\sigma^{2j-1}x^{k-j-2b}z) \ \textit{if}\ \sigma^2 = -(2k+1),$$

and

$$D(\sigma^{2k}z) = A_{0,0,k}D_\sigma(z) + \sum_{j=1}^{k}\sum_{b=0}^{[\frac{k-j+1}{2}]} A_{j,b,k}D(\sigma^{2j-2}x^{k-j-2b+1}z) \qquad \textit{if}\quad \sigma^2 = -2k,$$

and furthermore $A_{1,0,k} = 0$.

By blowing up one more time we get relations not involving $D_\sigma(z)$ but with degrees increased by 2. It is then a formal (non gauge-theoretic argument) to prove the main structure theorems for $\mathbf{D}$. These formalities are in the spirit the next two sections where we outline how to prove the general blowup formula. There we will determine the formal power series

$$B(x,t) = \sum_{k=0}^{\infty} B_k(x)\frac{t^k}{k!}$$

and

$$S(x,t)=\sum_{k=0}^{\infty} S_k(x)\frac{t^k}{k!}.$$

This will amount to solving an ordinary differential equation.

2.4. The ODE

Note:

$$\begin{aligned}\frac{d^n}{dt^n}\hat{D}(\exp(te)\,z) &= \hat{D}(\,z\sum e^{k+n}\frac{t^k}{k!}) = D(B_{k+n}(x)\frac{t^k}{k!}\,z)\\ &= \frac{d^n}{dt^n}D(B(x,t)\,z) = D(B^{(n)}(x,t)\,z)\end{aligned} \tag{2}$$

where the last differentiation is with respect to t. On $\bar{X}=X\#2\overline{\mathbf{CP}}^2$, we get $\bar{D}(\exp(t_1e_1+t_2e_2)z)=D(B(x,t_1)\,B(x,t_2)z)$. Now apply Corollary 2.2.2 to $e_1-e_2\in H_2(\bar{X};\mathbf{Z})$. Since for any $t\in\mathbf{R}$ the class $te_1+te_2\in\langle e_1-e_2\rangle^{\perp}$, we have the equation

$$\begin{aligned}&\bar{D}(\exp(te_1+te_2)\,(e_1-e_2)^4\,z)+4\,\bar{D}(x\,\exp(te_1+te_2)\,(e_1-e_2)^2\,z)\\ &\qquad\qquad +4\,\bar{D}(\exp(te_1+te_2)\,z)=0\end{aligned} \tag{3}$$

But, for example,

$$e_1^4\,\exp(te_1+te_2)=\left(\sum e_1^{k+4}\frac{t^k}{k!}\right)\left(\sum e_2^k\frac{t^k}{k!}\right)=\frac{d^4}{dt^4}(\exp(te_1))\,\exp(te_2)$$

Arguing similarly and using (2)

$$\begin{aligned}&\bar{D}(\exp(te_1+te_2)\,(e_1-e_2)^4\,z)\\ &= D((2\,B^{(4)}(x,t)\,B(x,t)-8\,B'''(x,t)\,B'(x,t)+6\,(B''(x,t))^2)\,z)\\ &= 2\,D((B^{(4)}\,B-4\,B'''\,B'+3\,(B'')^2)\,z)\end{aligned}$$

where $B=B(x,t)$. Completing the expansion of (3) we get

$$2\,D((B^{(4)}\,B-4\,B'''\,B'+3\,(B'')^2+4x\,(B''B-(B')^2)+2\,B^2)\,z)=0$$

for all $z\in\mathbf{A}(X)$. This means that the expression

$$B^{(4)}\,B-4\,B'''\,B'+3\,(B'')^2+4x\,(B''B-(B')^2)+2\,B^2$$

lies in the kernel of $D:\mathbf{A}(X)\to\mathbf{R}$.

Thus the "blowup function" $B(x,t)$ satisfies the differential equation

$$B^{(4)}\,B-4\,B'''\,B'+3\,(B'')^2+4x\,(B''B-(B')^2)+2\,B^2=0$$

modulo the kernel of D. Of course, the fact that this equation holds only modulo the kernel of D is really no constraint, since our interest in $B(x,t)$ comes from the equation $\hat{D}(\exp(te)z)=D(B(x,t)z)$.

Exercise 21. Modulo the kernel of D, the logarithm $f(t)$ of $B(x,t)$ satisfies the differential equation

$$f^{(4)}+6\,(f'')^2+4xf''+2=0 \tag{4}$$

with the initial conditions $f(0)=f'(0)=f''(0)=f'''(0)=0$. (Hint: Let $B=\exp(f(t))$. The initial conditions follow from Exercise 15.)

2.5. Solving the ODE

To solve the differential equation (4), set $u = f''$ and (4) becomes

$$u'' + 6u^2 + 4xu + 2 = 0 \tag{5}$$

with initial conditions $u(0) = u'(0) = 0$. This is a "standard" non-linear ODE whose solution is an elliptic function.

Exercise 22. Equation 5 is equivalent to

$$(u')^2 = -4u^3 - 4xu^2 - 4u \tag{6}$$

Exercise 23. Equation 6 is equivalent to

$$(y')^2 = 4y^3 - g_2 y - g_3 \quad \text{where} \quad g_2 = 4\,(\frac{x^2}{3} - 1) \text{ and } g_3 = \frac{8x^3 - 36x}{27}.$$

This is the equation which defines the Weierstrass $\wp$-function. In fact, if we rewrite (23) as

$$\frac{dt}{dy} = \frac{1}{\sqrt{4y^3 - g_2 y - g_3}}$$

then

$$t = \int_y^\infty \frac{ds}{\sqrt{4s^3 - g_2 s - g_3}} = \wp^{-1}(y) \tag{7}$$

and we see that for arbitrary constants c, $y = \wp(t + c)$ gives all solutions to (23), and so $f'' = u = -(\wp(t + c) + \frac{x}{3})$ is the general solution of (6).

Exercise 24. What does the integral (7) have to do with the ellipse?

The roots of the cubic equation

$$4s^3 - g_2 s - g_3 = 0$$

are

$$e_1 = \frac{x}{6} + \frac{\sqrt{x^2 - 4}}{2}, \quad e_2 = \frac{x}{6} - \frac{\sqrt{x^2 - 4}}{2}, \quad e_3 = -\frac{x}{3} \tag{8}$$

where we have followed standard notation (cf. [**Ak**]). These correspond to the half-periods $\omega_i = \wp^{-1}(e_i)$ of the $\wp$-function. The initial condition $f''(0) = 0$ implies that $\wp(c) = -\frac{x}{3} = e_3$; so $c = \omega_3 + 2\varpi$, where $2\varpi = 2m_1\omega_1 + 2m_3\omega_3$, with m_1, $m_3 \in \mathbf{Z}$, is an arbitrary period. (Note that the initial condition $f'''(0) = 0$ follows because the half-periods are zeros of $\wp'$.) The Weierstrass zeta-function satisfies $\zeta' = -\wp$; thus $f'(t) = \zeta(t + \omega_3 + 2\varpi) - \frac{tx}{3} + a$. The constant a is determined by the initial condition $f'(0) = 0$; $a = -\zeta(\omega_3 + 2\varpi)$. Since the logarithmic derivative of the Weierstrass sigma-function is ζ, integrating one more time gives $f(t) = \log\sigma(t + \omega_3 + 2\varpi) - t\zeta(\omega_3 + 2\varpi) - \frac{t^2 x}{6} + b$, and the initial condition $f(0) = 0$ shows that $b = -\log\sigma(\omega_3 + 2\varpi)$. Thus

$$B(x,t) = e^{f(t)} = e^{-\frac{t^2 x}{6}} e^{-t\zeta(\omega_3 + 2\varpi)} \frac{\sigma(t + \omega_3 + 2\varpi)}{\sigma(\omega_3 + 2\varpi)}.$$

For $\omega = \omega_1$ or ω_3 and $\eta = \zeta(\omega)$ we have the formulas

$$\zeta(u + 2m\omega) = 2m\eta + \zeta(u)\,, \qquad \sigma(u + 2m\omega) = (-1)^m e^{2\eta(mu + m^2\omega)} \sigma(u)$$

(which follow easily from [**Ak**, p.199]). Using them, our formula for $B(x,t)$ becomes

$$B(x,t) = e^{-\frac{t^2 x}{6}} e^{-\eta_3 t} \frac{\sigma(t+\omega_3)}{\sigma(\omega_3)}.$$

The above addition formula for the sigma-function implies that

$$\sigma(t+\omega_3) = \sigma((t-\omega_3)+2\omega_3) = -e^{2\eta_3 t}\sigma(t-\omega_3).$$

Thus

$$B(x,t) = -e^{-\frac{t^2 x}{6}} e^{\eta_3 t} \frac{\sigma(t-\omega_3)}{\sigma(\omega_3)} = e^{-\frac{t^2 x}{6}} \sigma_3(t),$$

the last equality by the definition of the quasi-periodic function σ_3. In conclusion,

Theorem 2.5.1. ([**FS8**]) *Modulo the kernel of D, the blowup function $B(x,t)$ is given by the formula*

$$B(x,t) = e^{-\frac{t^2 x}{6}} \sigma_3(t). \qquad \square$$

The indexing of the Weierstrass functions σ_i depends on the ordering of the roots e_i of the equation $4s^3 - g_2 s - g_3 = 0$; the sigma-function we are using corresponds to the root $-\frac{x}{3}$.

These ideas also show

Theorem 2.5.2. ([**FS8**]) *Modulo the kernel of D, the blowup function $S(x,t)$ is given by the formula*

$$S(x,t) = e^{-\frac{t^2 x}{6}} \sigma(t).$$

2.6. The simple-type condition

An elliptic function is a doubly-periodic function, so it is interesting to see what happens when one of the periods goes to infinity. This gives a degenerate case of the associated Weierstrass functions. The squares k^2, k'^2 of the modulus and complementary modulus of our Weierstrass functions are given by

$$k^2 = \frac{x-\sqrt{x^2-4}}{x+\sqrt{x^2-4}} \qquad k'^2 = \frac{2\sqrt{x^2-4}}{x+\sqrt{x^2-4}}.$$

(All the formulas involving elliptic functions can be found in [**Ak**].) These degenerate when $x=2$ and then $k^2 = 1$ and $k'^2 = 0$. The corresponding complete elliptic integrals of the first kind are

$$\begin{aligned} K &= \int_0^1 \frac{ds}{\sqrt{(1-s^2)(1-k^2s^2)}} = \int_0^1 \frac{ds}{1-s^2} \\ K' &= \int_0^1 \frac{ds}{\sqrt{(1-s^2)(1-k'^2s^2)}} = \int_0^1 \frac{ds}{\sqrt{1-s^2}} \end{aligned}$$

Thus $K = \infty$ and $K' = \frac{\pi}{2}$ when $x = 2$. Also, when $x = 2$ we have $g_2 = \frac{4}{3}$ and $g_3 = -\frac{8}{27}$; so the roots of $4s^3 - g_2 s - g_3 = 0$ are $e_1 = e_2 = \frac{1}{3}$ and $e_3 = -\frac{2}{3}$. This means that when $x = 2$ the basic periods are

$$\omega_1 = \frac{K}{\sqrt{e_1 - e_3}} = K = \infty \qquad \omega_3 = \frac{iK'}{\sqrt{e_1 - e_3}} = iK' = \frac{i\pi}{2}.$$

In this situation,

$$\sigma(t) = \frac{2\omega_3}{\pi} e^{\frac{1}{6}(\frac{\pi t}{2\omega_3})^2} \sin\frac{\pi t}{\omega_3} = e^{-\frac{t^2}{6}} \sinh t$$

and

$$\wp(t) = -\frac{1}{3}(\frac{\pi}{2\omega_3})^2 + (\frac{\pi}{2\omega_3})^2 \frac{1}{\sin^2(\frac{\pi t}{2\omega_3})} = \frac{1}{3} + \frac{1}{\sinh^2 t}.$$

So

$$\sigma_3(t) = \sigma(t)\sqrt{\wp(t) - e_3} = e^{-\frac{t^2}{6}} \sinh t \sqrt{1 + \frac{1}{\sinh^2 t}} = e^{-\frac{t^2}{6}} \cosh t.$$

Recall the *Donaldson series* of X given by

$$\mathbf{D}(\alpha) = D((1 + \frac{x}{2}) \exp(\alpha))$$

for all $\alpha \in H_2(X)$. When $x = 2$,

$$D((1 + \frac{x}{2}) z x) = 2 D((1 + \frac{x}{2}) z),$$

which is equivalent to the simple-type condition $D(x^2 z) = 4 D(z)$ for all $z \in \mathbf{A}(X)$.

Theorem 2.6.1. *([**FS8**]) If X has simple type, the Donaldson series of $\hat{X} = X \# \overline{\mathbf{CP}}^2$ is*

$$\hat{\mathbf{D}} = \mathbf{D} \cdot e^{-\frac{E^2}{2}} \cosh E$$

where E is the form dual to the exceptional class e, i.e. $E(z) = e \cdot z$ for all $z \in H_2(\hat{X})$. Also

$$\hat{\mathbf{D}}_e = -\mathbf{D} \cdot e^{-\frac{E^2}{2}} \sinh E.$$

Proof. For $\alpha \in H_2(X)$ we calculate

$$\begin{aligned} \hat{\mathbf{D}}(\alpha + te) &= \hat{D}((1 + \frac{x}{2}) \exp(\alpha) \exp(te)) = D((1 + \frac{x}{2}) \exp(\alpha) B(x,t)) \\ &= D((1 + \frac{x}{2}) \exp(\alpha) e^{-\frac{t^2 x}{6}} e^{-\frac{t^2}{6}} \cosh t) \end{aligned}$$

The simple type condition implies that $D((1 + \frac{x}{2}) e^{-\frac{t^2 x}{6}}) = D((1 + \frac{x}{2}) e^{-\frac{t^2}{3}})$. Hence

$$\hat{\mathbf{D}}(\alpha + te) = \mathbf{D}(\alpha) e^{-\frac{t^2}{2}} \cosh t = \mathbf{D}(\alpha)(e^{-\frac{E^2}{2}} \cosh E)(te)$$

as desired. The formula for $\hat{\mathbf{D}}_e$ follows similarly since $\sinh(E)(te) = -\sinh(t)$.

Recall that a 4-manifold X is said to have *c-simple type* if, for $c \in H_2(X; \mathbf{Z})$, $D_c(x^2 z) = 4 D_c(z)$ for all $z \in \mathbf{A}(X)$. It is shown in [**FS9**], and also by Kronheimer and Mrowka, that if X has simple type, then it has c-simple type for all $c \in H_2(X; \mathbf{Z})$. As above we have,

Theorem 2.6.2. *([**FS8**])If X has c-simple type,*

$$\begin{aligned}\hat{\mathbf{D}}_c &= \mathbf{D}_c \cdot e^{-\frac{E^2}{2}} \cosh E\,. \\ \hat{\mathbf{D}}_{c+e} &= -\mathbf{D}_c \cdot e^{-\frac{E^2}{2}} \sinh E\,. \qquad \square\end{aligned}$$

Exercise 25. If X has simple type, determine the basic classes of $X\#\overline{\mathbf{CP}}^2$.

All of this from relations involving spheres with self-intersection −2. There is another application that we will discuss in the next lecture.

LECTURE 3
Simple-Type Criteria and Elliptic Surfaces

3.1. Manifolds with big diffeomorphism group

Using Theorem 2.1.1 we show that for many manifolds the homogeneous parts of D_X are related, show these manifolds have simple-type. For a 4-manifold X and a class $\kappa \in H_2(X;\mathbf{Z})$, let $\mathrm{Diff}_\kappa(X)$ be the group of orientation-preserving diffeomorphisms f of X which satisfy $f_*(\kappa) = \kappa$. Also, let $\mathrm{Aut}(X)$, be the group of automorphisms of $H_2(X;\mathbf{Z})$ which preserve the intersection form Q. Then X is said to have a *big diffeomorphism group* with respect to κ if the image of $\mathrm{Diff}_\kappa(X)$ in $\mathrm{Aut}(X)$ has finite index. For example, the simply connected minimal elliptic surfaces with $p_g \geq 1$ have a big diffeomorphism group with respect to their canonical class [**FM2**]. It follows from the assumption of big diffeomorphism group with respect to κ that for each d, the degree d homogeneous part $D^{(d)}_{X,c}$ of the Donaldson invariants $D_{X,c}$ and $D_{X,c}(\frac{x}{2})$ (and hence $\hat{D}_{X,c}$) are polynomials in the intersection form Q and the class κ when viewed as linear maps $\mathrm{Sym}_*(H_2(X)) \to \mathbf{R}$. If $\frac{1}{2}(1 + b^+_X) \equiv 0 \pmod 2$ we can then write

$$\begin{aligned}\frac{1}{(2d)!}\hat{D}^{(2d)}_X &= c_0\frac{Q^d}{2^d d!} + c_2\frac{Q^{d-1}}{2^{d-1}(d-1)!}\frac{\kappa^2}{2!} + \dots \\ &+ c_{2t}\frac{Q^{d-t}}{2^{d-t}(d-t)!}\frac{\kappa^{2t}}{(2t)!} + \dots + c_{2d}\frac{\kappa^{2d}}{(2d)!}\end{aligned}$$

and if $\frac{1}{2}(1 + b^+_X) \equiv 1 \pmod 2$ we can write

$$\begin{aligned}\frac{1}{(2d+1)!}\hat{D}^{(2d+1)}_X &= c_1\frac{Q^d}{2^d d!}\kappa + \dots \\ &+ c_{2t+1}\frac{Q^{d-t}}{2^{d-t}(d-t)!}\frac{\kappa^{2t+1}}{(2t+1)!} + \dots + c_{2d+1}\frac{\kappa^{2d+1}}{(2d+1)!}.\end{aligned}$$

Our next Proposition states that if X contains an embedded 2-sphere of self-intersection -2 which is orthogonal to κ, the coefficients c_j are independent of the homogeneous degree. Related results were first observed by Peter Kronheimer (unpublished).

Proposition 3.1.1. ([**FS9**]) *Suppose that $\sigma \in H_2(X;\mathbf{Z})$ is represented by an embedded 2-sphere, and suppose also that $\sigma^2 = -2$ and $\sigma \cdot \kappa = 0$. Then the coefficients c_j above are independent of the homogeneous degree d of $\hat{D}_X^{(d)}$. Similarly, if we express $\hat{D}_{X,\omega}$ as a polynomial in Q and κ with coefficients c'_j, and if also $\omega \cdot \sigma \equiv 0$* (mod 2), *then the c'_j are independent of the degree d of $\hat{D}_{X,\omega}^{(d)}$.*

Proof. To fix notation, assume that $\frac{1}{2}(1+b_X^+) \equiv 0 \pmod 4$. Write

$$\frac{1}{(2d-2)!}\hat{D}_{X,\sigma}^{(2d-2)} = c'_0\frac{Q^{d-1}}{2^{d-1}(d-1)!} + c'_2\frac{Q^{d-2}}{2^{d-2}(d-2)!}\frac{\kappa^2}{2!} + \dots$$

Consider an $\alpha \in H_2(X)$ which satisfies $\alpha\cdot\sigma = \alpha\cdot\kappa = 0$. Then

$$\begin{aligned}
\hat{D}_X(\alpha^{2d-2}\sigma^2) &= (2d)!c_0\frac{Q^d}{2^d d!}(\alpha^{2d-2}\sigma^2)\\
&= \frac{(2d)!c_0}{2^d d!}\frac{d\cdot 2!(2d-2)!}{(2d)!}Q^{d-1}(\alpha^{2d-2})Q(\sigma^2)\\
&= \frac{-2c_0(2d-2)!}{2^{d-1}(d-1)!}Q^{d-1}(\alpha^{2d-2})
\end{aligned}$$

But, by Theorem 2.1.2

$$\hat{D}_X(\alpha^{2d-2}\sigma^2) = 2\,\hat{D}_{X,\sigma}(\alpha^{2d-2}) = 2(2d-2)!\frac{c'_0}{2^{d-1}(d-1)!}Q^{d-1}(\alpha_{2d-2}).$$

Thus $c'_0 = -c_0$.

Next take $\beta \in H_2(X)$ with $\beta^2 = 0$, $\beta\cdot\kappa = 1$, and $\beta\cdot\sigma = \beta\cdot\alpha = 0$. Then for $j \geq 1$

$$\begin{aligned}
\hat{D}_X(\alpha^{2d-2j-2}\beta^{2j}\sigma^2) &= (2d)!\frac{c_{2j}}{2^{d-j}(d-j)!}Q^{d-j}\cdot\frac{\kappa^{2j}}{2j!}(\alpha^{2d-2j-2}\beta^{2j}\sigma^2)\\
&= -c_{2j}\frac{(2d-2j-2)!}{2^{d-j-2}(d-j-1)!}Q^{d-j-1}(\alpha^{2d-2j-2})\\
= 2\,\hat{D}_{X,\sigma}(\alpha^{2d-2j}\beta^2) &= c'_{2j}\frac{(2d-2j-2)!}{2^{d-j-2}(d-j-1)!}Q^{d-j-1}(\alpha^{2d-2j-2}).
\end{aligned}$$

Thus $c'_{2j} = -c_{2j}$ for all $j \geq 0$.

We now perform the same procedure on the $SO(3)$-Donaldson invariants $D_{X,\sigma}$. Write

$$\frac{1}{(2d-4)!}\hat{D}_X^{(2d-4)} = c''_0\frac{Q^{d-2}}{2^{d-2}(d-2)!} + c''_2\frac{Q^{d-3}}{2^{d-3}(d-3)!}\frac{\kappa^2}{2!} + \dots$$

Note that Theorem 2.1.2 implies that for all $z \in \mathbf{A}(\sigma^\perp)$,

$$\hat{D}_{X,\sigma}(\sigma^2 z) = 2\,\hat{D}_{X,2\sigma}(z) = 2(-1)^{\sigma^2}\hat{D}_X(z) = 2\,\hat{D}_X(z)\,.$$

Using the expansion of $\hat{D}_X^{(2d-4)}$ and applying Theorem 2.1.2, we get

$$\begin{aligned}
\hat{D}_{X,\sigma}(\alpha^{2d-4}\sigma^2) &= (2d-2)!\frac{c'_0}{2^{d-1}(d-1)!}Q^{d-1}(\alpha^{2d-4}\sigma^2)\\
&= -c'_0\frac{(2d-4)!}{2^{d-3}(d-2)!}Q^{d-2}(\alpha^{2d-4})\\
= 2\,\hat{D}_X(\alpha^{2d-4}) &= 2(2d-4)!\frac{c''_0}{2^{d-2}(d-2)!}Q^{d-2}(\alpha^{2d-4}).
\end{aligned}$$

Thus $c_0' = -c_0''$ and so $c_0 = c_0''$. In a similar fashion we get that $c_{2j} = c_{2j}''$ for all j.

Finally consider $\hat{D}_X^{(2d-2)} = D_X^{(2d)}(\frac{x}{2})$. Because X has a big diffeomorphism group we can write

$$\frac{1}{(2d-2)!}\hat{D}_X^{(2d-2)} = \hat{c}_0\frac{Q^{d-1}}{2^{d-1}(d-1)!} + \hat{c}_2\frac{Q^{d-2}}{2^{d-2}(d-2)!}\frac{\kappa^2}{2!} + \dots$$

Proposition 2.2.2 implies that

$$D_X(\alpha^{2d}\sigma^4) = -8\, D_X(\alpha^{2d}\sigma^2\frac{x}{2}) - 4\, D_X(\alpha^{2d}) .$$

Expanding as above, and using the fact that $c_0'' = c_0$, we get $\hat{c}_0 = c_0$, and continuing as above, $\hat{c}_i = c_i$ for all i. This completes the proof in the $SU(2)$ case with $\frac{1}{2}(1+b_X^+) \equiv 0 \pmod 4$. A similar proof suffices when $\frac{1}{2}(1+b_X^+) \not\equiv 0 \pmod 4$, and the same proof also works in the $SO(3)$ case.

3.2. A simple-type criteria

A related argument shows that these manifolds have simple-type.

Theorem 3.2.1. *([FS9]) Let X be a simply connected 4-manifold which has a big diffeomorphism group with respect to a class $\kappa \in H_2(X;\mathbf{Z})$, and let $\omega \in H_2(X;\mathbf{Z})$. Suppose that $\sigma \in H_2(X;\mathbf{Z})$ is represented by an embedded 2-sphere of square -2 such that $\sigma\cdot\kappa = 0$, and $\sigma\cdot\omega \equiv 0 \pmod 2$. Then X has ω-simple type. If $\omega^2 + \frac{1}{2}(1+b^+) \equiv 0 \pmod 2$ then*

$$\mathbf{D}_{X,\omega} = \exp(Q/2)\sum c_{2i,\omega}\frac{\kappa^{2i}}{(2i)!} ,$$

and if $\omega^2 + \frac{1}{2}(1+b^+) \equiv 1 \pmod 2$ then

$$\mathbf{D}_{X,\omega} = \exp(Q/2)\sum c_{2i+1,\omega}\frac{\kappa^{2i+1}}{(2i+1)!} .$$

Proof. For simplicity of notation, consider the $SU(2)$ case $\omega = 0$ with $\frac{1}{2}(1+b_X^+) \equiv 0 \pmod 4$ as above. Then Proposition 3.1.1 shows that for each $d \equiv \frac{1}{2}(1+b^+) \pmod 4$ the homogeneous invariants $D_X^{(2d+4)}(\frac{x}{2}) = \hat{D}_X^{(2d+2)}$ and $D_X^{(2d)}$ share the same coefficients in that

$$\frac{1}{2d!}D_X^{(2d)} = \sum_{i=0}^{d} c_{2i}\frac{Q^{d-i}}{2^{d-i}(d-i)!}\frac{\kappa^{2i}}{(2i)!}$$

$$\frac{1}{(2d+2)!}D_X^{(2d+4)}(\frac{x}{2}) = \sum_{i=0}^{d+1} c_{2i}\frac{Q^{d+1-i}}{2^{d+1-i}(d+1-i)!}\frac{\kappa^{2i}}{(2i)!}$$

The technique of Proposition 3.1.1 also shows that

$$\frac{1}{(2d+4)!}D_X^{(2d+8)}(\frac{x^2}{4}) = \sum_{i=0}^{d+2} c_{2i}\frac{Q^{d+2-i}}{2^{d+2-i}(d+2-i)!}\frac{\kappa^{2i}}{(2i)!} ,$$

independent of degree. Thus also,

$$\frac{1}{2d!}D_X^{(2d+4)}(\frac{x^2}{4}) = \sum_{i=0}^{d} c_{2i}\frac{Q^{d-i}}{2^{d-i}(d-i)!}\frac{\kappa^{2i}}{(2i)!} = \frac{1}{(2d)!}D_X^{(2d)} ;$$

so X has simple type. The other cases are similar.

Theorem 3.2.1 can be used to show a large class of manifolds have simple-type. After all, all the proofs that manifolds have big diffeomorphism group use the fact that there are many -2 curves which determine diffeomorphisms which generate the diffeomorphism group of X. In the algebraic geometric setting, these arise as vanishing cycles for deformations of the surfaces in question. So for example, all elliptic surfaces (with $\chi \geq 2$), complete intersections, Moisheson and Salvetti surfaces have simple-type. Further, this argument shows that whenever X has a -2 curve S such that $\mathbf{D}(z,[S]^r) = \exp(Q/2)(z,[S]^r)$ for all $z \in \mathbf{A}([S])^{\perp}$, then X has simple-type. There is no known 4-manifold which violates this criteria.

Problem 18. Find an example of a complex surface with no deformations and determine if it has simple-type.

3.3. The Donaldson series for the $K3$ surface

The following result has been "known" for some time; here's a proof.

Corollary 3.3.1. *The $K3$ surface, $E(2)$, has c-simple type for all $c \in H_2(E(2);\mathbf{Z})$ and*

$$\mathbf{D}_{E(2),c} = (-1)^{\frac{c^2}{2}} \exp(Q/2)\,.$$

Proof. The diffeomorphism group Diff($E(2)$) acts on $H_2(E(2);\mathbf{Z}_2)$ with exactly 3 orbits, namely $\{0\}$, $\{\alpha \neq 0 | \alpha^2 \equiv 0 \pmod 4\}$, and $\{\alpha | \alpha^2 \equiv 2 \pmod 4\}$. These are represented by 0, f, and s_2, the class of a section. Thus, keeping in mind the rule $D_{c'} = (-1)^{(\frac{c-c'}{2})^2} D_c$ for $c \equiv c' \pmod 2$, we may assume that c is one of these 3 classes.

In case $c = 0$ or s_2, we let $\sigma = s_2$. If $c = f$ then there is a different elliptic fibration whose section $\sigma = s_2'$ satisfies $s_2' \cdot f = 0$. Now $E(2)$ has a big diffeomorphism group (with respect to $\kappa = 0$); so we may apply Theorem 3.2.1 to see first that $E(2)$ has c-simple type for all c, and second, that $\mathbf{D}_{E(2),c} = a_c \exp(Q/2)$ for some constant a_c. For $c^2 \equiv 2 \pmod 4$ it is known that $D_{E(2),c}(1) = -1$, and for $c^2 \equiv 0 \pmod 4$ it is known that $D^{(14)}_{E(2),c} = (14)!(Q^7/(2^7 \cdot 7!)$ [**DK**]. It follows that for any c, $a_c = (-1)^{\frac{1}{2}c^2}$. The equation for $\mathbf{D}_{E(2),c}$ follows.

3.4. Another simple-type criteria

Another useful criteria for determining simple-type is the following Corollary which follows from the fact that the Brieskorn manifold $B(2,3,7)$ (= Milnor fiber of the $(2,3,7)$ singularity $\Sigma(2,3,7)$) resides in the $K3$ surface and that the homology sphere $\Sigma(2,3,7)$ has sparse Floer homology.

Corollary 3.4.1. *([**FS9**]) Any simply connected 4-manifold (with $b^+ \geq 3$ and odd) which contains the Brieskorn manifold $B(2,3,7)$ has c-simple type for all c.*

3.5. The Donaldson series for elliptic surfaces

We are now in a position to compute the Donaldson series for the elliptic surfaces without multiple fibers. This is done by inductively constructing the $E(n)$ from

$E(n-1)\#\overline{\mathbf{CP}}^2$ and using the blowup formula. Here's where a little topology helps out. It turns out that $E(n)$, $n \geq 2$, can be decomposed as

$$E(n) = B(2,3,11) \cup W(n).$$

That is,

Exercise 26. The Milnor fiber $B(2,3,11)$ embeds in $E(n)$, $n \geq 2$. (Hint: The Milnor fibers are nested: $B(p_1, p_2, \ldots, p_r) \subset B(p_1', p_2', \ldots, p_r')$ whenever $p_j \leq p_j'$ for all j and $E(n) = B(2,3,6n-1) \cup G_n$, where G_n is a so-called Gompf nucleus [**G1**].)

Now $\partial B(2,3,11)$ bounds another interesting manifold, $C(2,3,11)$ obtained as the union $C(2,3,11) = B(2,3,5) \cup D$, where D is constructed by attaching one 2-handle to $\Sigma(2,3,5)$ This follows since $-\Sigma(2,3,5)$ is -1 surgery on the left hand trefoil knot and $-\Sigma(2,3,11)$ is $-\frac{1}{2}$ surgery on the same knot.

Exercise 27. $C(2,3,11) \cup W(n) = E(n-1)\#\overline{\mathbf{CP}}^2$.

This is our inductive construction: replace $C(2,3,11)$ in $E(n-1)\#\overline{\mathbf{CP}}^2$ by $B(2,3,11)$ to obtain $E(n)$.

We are after the following computation:

Theorem 3.5.1. *([**FS9**]) The Donaldson series of the elliptic surfaces $E(n)$ are given by*

$$\mathbf{D}_{E(n),c} = (-1)^{\frac{1}{2}(c^2+(n-2)c\cdot f)} \exp(Q/2)\sinh^{n-2}(f) \qquad \text{if} \quad c\cdot f \equiv 0 \pmod 2,$$
$$\mathbf{D}_{E(n),c} = (-1)^{\frac{1}{2}(c^2+(n-2)c\cdot f)} \exp(Q/2)\cosh^{n-2}(f) \qquad \text{if} \quad c\cdot f \equiv 1 \pmod 2.$$

where $f \in H_2(E(n);\mathbf{Z})$ is the homology class of the fiber.

This result has also been obtained by Paulo Lisca [**Li**] and Kronheimer-Mrowka [**KM1**].

Exercise 28. Determine the basic classes for $E(n)$.

The idea of the proof is simple; suppose $c \cdot f \equiv 0 \pmod 2$. Since elliptic surfaces have a big diffeomorphism group with respect to their fiber and simple-type, it follows by Theorem 3.1.1 that $\mathbf{D}_{E(n)} = \exp(Q/2)\mathbf{K}$ where $\mathbf{K}$ is a formal power series in f. Assume inductively that

$$\mathbf{D}_{E(n),c} = (-1)^{\frac{1}{2}(c^2+(n-2)c\cdot f)} \exp(Q/2)\sinh^{n-2}(f)$$

$K3 = E(2)$ starts the induction. The blow-up formula says that

$$\mathbf{D}_{E(n)\#\overline{\mathbf{CP}}^2,c} = (-1)^{\frac{1}{2}(c^2+(n-2)c\cdot f)} \exp(Q/2)\sinh^{n-2}(f)\cosh(e)$$

But $C(2,3,11) \cup W(n) = E(n-1)\#\overline{\mathbf{CP}}^2$. The thing to show is that by replacing $C(2,3,11)$ by $B(2,3,11)$ replaces $\cosh(e)$ by $\sinh(f)$. Since the Floer homology, $HF_*(\Sigma(2,3,11))$ is a copy of $\mathbf{Z}$ in odd dimensions and vanishes in even dimensions [**FS3**], this becomes a fun exercise. A technical point here is that when you stretch the neck of $E(n-1)\#\overline{\mathbf{CP}}^2$ along $\Sigma(2,3,11)$, the trivial connection doesn't occur as asymptotic value –guaranteeing the fact that reducible connections in the negative definite manifold $C(2,3,11)$ do not complicate the picture. The proof here is interesting, and is again typical of many of our arguments in [**FS9**].

Lemma 3.5.2. *Suppose that* $X = X_1 \cup X_2$ *with* $\partial X_1 = -\partial X_2 = \Sigma(2,3,11)$, *and assume that* $b^+_{X_1} = 0$ *and* $b^+_{X_2} > 0$. *Let* $c_i \in H_2(X_i;\mathbf{Z})$, $i = 1,2$. *If* u_i *has degree* d_i *in* $\mathbf{A}(X_i)$ *and* $d_1 \not\equiv c_1^2 \pmod 2$ *then*

$$\hat{D}_{X,c_1+c_2}(u_1u_2) = \langle D_{X_1,c_1}[\beta](u_1), D_{X_2,c_2}[\beta](u_2)\rangle$$

Proof. Contributions to $D_{X,c_1+c_2}(u_1u_2)$ arise from products of compactified based moduli spaces $\tilde{\mathcal{M}}^c_{X_1}[\zeta] \times \tilde{\mathcal{M}}^c_{X_2}[\zeta]$, divided out by the diagonal $SO(3)$, where ζ denotes α, β, or ϑ. (The appropriate compactification is due to Taubes [**T4**].) We are concerned with the case $\zeta = \vartheta$. Since the stabilizer of ϑ is 3 (> 0)-dimensional, a counting argument shows that the divisor corresponding to u_1 does not intersect $\mathcal{M}^c_{X_1}[\vartheta]$ transversely and that the compactified based moduli space $\tilde{\mathcal{M}}^c_{X_1}[\vartheta]$ contains reducible nontrivial orbits. Each of these reducible orbits is a 2-sphere. The class $\mu(u_1)$ lifts to an $SO(3)$ equivariant cohomology class $\tilde{\mu}(u_1) \in H^{2d_1}_{SO(3)}(\tilde{\mathcal{M}}_{X_1}[\vartheta];\mathbf{R})$ which extends to the compactification, $\tilde{\mu}(u_1) \in H^{2d_1}_{SO(3)}(\tilde{\mathcal{M}}^c_{X_1}[\vartheta];\mathbf{R})$ (see[**T4**] and [**FS9**]). Since each (based) connection in $\tilde{\mathcal{M}}^c_{X_1}[\vartheta]$ has the same asymptotic value, which we can identify with $1 \in SO(3)$, there is an $SO(3)$-equivariant push-forward map

$$\partial_* : H^*_{SO(3)}(\tilde{\mathcal{M}}^c_{X_1}[\vartheta];\mathbf{R}) \to H^*_{SO(3)}(\{1\};\mathbf{R})$$

given by integration along the fiber. The fiber dimension of the map $\partial : \tilde{\mathcal{M}}^c_{X_1} \to SO(3)$ is equal to $\dim \tilde{\mathcal{M}}^c_{X_1}[\vartheta] \equiv -2c_1^2 - 8k$ for some k. There is similarly a pullback map

$$\partial^* : H^*_{SO(3)}(\{1\};\mathbf{R}) \to H^*_{SO(3)}(\tilde{\mathcal{M}}^c_{X_2}[\vartheta];\mathbf{R}).$$

Furthermore, since $SO(3)$ acts freely on $\tilde{\mathcal{M}}^c_{X_2}[\vartheta]$ there is an isomorphism

$$\pi^* : H^*(\mathcal{M}^c_{X_2}[\vartheta];\mathbf{R}) \cong H^*_{SO(3)}(\tilde{\mathcal{M}}^c_{X_2}[\vartheta];\mathbf{R}).$$

Then the corresponding contribution to the Donaldson invariant is obtained from the pairing $\langle \pi^*\partial^*\partial_*\tilde{\mu}(u_1) \cup \mu(u_2), [\mathcal{M}^c_{X_2}[\vartheta]]\rangle$. (Again, see [**T2, T3, T4, AB**] and [**FS9**] for this.) However, under the hypothesis of the lemma, $2d_1 + 2c_1^2 \equiv 2 \pmod 4$; so

$$\partial_*\tilde{\mu}(u_1) \in H^{2d_1+2c_1^2+8k}_{SO(3)}(\{1\};\mathbf{R}) \cong H^{2d_1+2c_1^2+8k}(BSO(3);\mathbf{R}) = 0.$$

This means that the boundary value ϑ does not contribute to the calculation of the invariant $D_{X,c_1+c_2}(u_1,u_2)$. The correct boundary value lives in $HF_j(\Sigma(2,3,11))$, where

$$j \equiv -(2c_1^2 + 3 + 2d_1) \equiv -(2+3) \pmod 4.$$

Thus it must be β.

LECTURE 4
Elementary Rational Blowdowns

4.1. Elementary rational blowdowns

There are two directions in which to generalize our basic relations Theorem 2.1.1 and Theorem 2.2.1. The first, as we have mentioned before, is to prove relations for spheres of arbitrary self-intersection. This is done in [**FS9**] and results in Theorem 2.3.1. This in turn implies the general structure theorems and adjunction formulas for the Donaldson invariants mentioned in the first lecture. Here's another fruitful generalization. Suppose S is a sphere with self-intersection -4 and let N $(=C_2))$ be a tubular neighborhood of S so that $X = X_0 \cup N$.

Exercise 29. ∂N bounds the Euler class -2 non-orientable S^1-bundle over $\mathbf{RP}^2$, a rational ball B_2.

Let X_2 be the result of replacing N by B_2.

Proposition 4.1.1. *For all $z \in \mathbf{A}(X_2)$,*

$$D_{X_2}(z) = D_X(z) - D_{X,[S]}(z).$$

Proof. We outline the proof. Let $L = \partial N$ and $\chi(L) = \{\zeta^0 = \vartheta, \zeta, \zeta^2, \zeta^3\}$ be the character variety of L. The only characters which extend across B_2 are ϑ and ζ^2. Thus $D_{X_2}(z) = \pm D_{X_0}[\vartheta](z) \pm D_{X_0}[\zeta](z)$. By stretching the neck of X along L, dimension counting shows that $D_{X_0}[\vartheta](z) = D_X(z)$ (for the dimension of $\mathcal{M}_N = -3$). To determine $D_{X_0}[\zeta](z)$, we do as in Theorem 2.1.1 and glue in the trivial bundle over N with a twist to obtain $D_{X_0}[\zeta](z) = \pm D_{X,[S]}(z)$. Again, a delicate (but important) point is to get the correct signs. For this we refer to [**FS10**]. Now the fun begins!

Consider $E(4)$ with S being the section. Since

$$\mathbf{D}_{E(4)} = \exp(\frac{Q}{2}) \sinh^2(F)$$

and

$$\mathbf{D}_{E(4),S} = (-1)^{\frac{S^2+2F\cdot S}{2}} \exp(\frac{Q}{2}) \cosh^2(F) = -\exp(\frac{Q}{2}) \cosh^2(F),$$

we see that on $E(4) \setminus C_2$

$$\mathbf{D}_{X_2} = \exp(\frac{Q}{2})(\sinh^2(F) + \cosh^2(F)) = \exp(\frac{Q}{2}) \cosh(2F)$$

Now $2F$ is not a cohomology class in X_2 since $Q(2f, [S]) \neq 0$. However, the appropriate class is $\kappa = \frac{1}{2}(4F + PD[S])$ (since we can add multiples of classes in B_2) so that

$$\mathbf{D}_{X_2} = \exp(\frac{Q}{2}) \cosh(\kappa)$$

with $\kappa^2 = +1$.

This is an interesting 4-manifold. First, it is not a blow-up, since we see no $\cosh(E)$ in its Donaldson series. Thus if it were an algebraic surface, it would be minimal. However, this surface violates the Noether inequality $c_1^2(X) \geq 2\chi - 6$, so it is not homeomorphic to any algebraic surface -but it has a non-trivial Donaldson series. This manifold, however, does indeed have a symplectic structure [**G2**] and, in fact, κ is the first Chern class of the symplectic structure.

Exercise 30. Any $E(n)$ has nine disjoint sections with self-intersection $-n$.

Let X_ℓ denote the result of blowing down ℓ sections in $E(4)$.

Exercise 31. $\mathbf{D}_{X_\ell} = 2^{\ell-1} \exp(\frac{Q}{2}) \cosh(\kappa_\ell)$ with $\kappa_\ell^2 = \ell$.

Exercise 32. If $\ell \geq 2$, then X_ℓ satisfies the Noether inequalities.

Advanced exercises are

Exercise 33. X_2 is diffeomorphic to the double branched cover of the octic in $\mathbf{CP}^2$.

Exercise 34. Show X_ℓ, $\ell = 2, 3, 4, 9$ are complex surfaces.

Problem 19. Determine if X_ℓ, $\ell = 5, 6, 7, 8$ are complex surfaces.

4.2. Logarithmic transform as rational blowdown

As mentioned earlier, Fintushel and I discovered rational blowdowns in order to compute the Donaldson series for the logarithmic transforms of the $E(n)$. Recall, that in $E(n)$ the tubular neighborhood of each of the torus fibers is a copy of $T^2 \times D^2 = S^1 \times (S^1 \times D^2)$. By a *log transform* on $E(n)$, we mean the result of removing this $T^2 \times D^2$ from $E(n)$ and regluing it by a diffeomorphism

$$\varphi : T^2 \times \partial D^2 \to T^2 \times \partial D^2.$$

The *multiplicity* of the surgery is the absolute value of the degree of

$$\mathrm{pr}_{\partial D^2} \circ \varphi : \mathrm{pt} \times \partial D^2 \to \partial D^2.$$

Let $E(n)_\varphi$ denote the result of this operation on $E(n)$. Note that multiplicity 0 is a possibility. It follows from Moishezon [**Mn1**], that if φ and φ' have the same multiplicity, there is a diffeomorphism, fixing the boundary, from $E(n)_\varphi$ to $E(n)_{\varphi'}$. What is needed here is that nearby there are cusp-fibers (cf. [**FS6**]), i.e. an immersed 2-sphere with self-intersection 0 and one positive double point. Let $E(n;p)$ denote any $E(n)_\varphi$ where the multiplicity of φ is p.

In $E(n;p)$ there is again a copy of the fiber F, but there is also a new torus fiber, the *multiple fiber.* Denote its homology class by f_p; so in $H_2(E(n;p); \mathbf{Z})$

we have $f = p\,f_p$. Continue this process on other fibers, but to insure that the resulting manifold is simply-connected we can take at most two log-transforms with multiplicities that are pairwise relatively prime. Let the orders be p and q and denote the result by $E(n;p,q)$.

The homology class f of the fiber of $E(n)$ can be represented by an immersed sphere with one positive double point (a nodal fiber). Blow-up this double point (i.e. take the proper transform of f) so that the class $f-2e_1$ (where e_1 is the homology class of the exceptional divisor) is represented by an embedded sphere with square -4. This is just the configuration C_2. Now the exceptional divisor intersects this sphere in two positive points. Blow-up one of these points, i.e. again take a proper transform. There results the homology classes $u_0 = f - 2e_1 - e_2$ and $u_1 = e_1 - e_2$ which is just the configuration C_3. Continuing in this fashion, C_p naturally embeds in $N\#_{p-1}\overline{\mathbf{CP}}^2 \subset E(n)\#_{p-1}\overline{\mathbf{CP}}^2$.

An important example of a rational blowdown is:

Theorem 4.2.1 ([**FS10**]). *The rational blowdown of the above configuration $C_p \subset E(n)\#(p-1)\overline{\mathbf{CP}}^2$ is diffeomorphic $E(n;p)$.*

When $p=2$, this was first observed by Gompf [**G3**]

We can now easily compute $\mathbf{D}_{E(2;2)}$ on $X \setminus C_2$.

$$\begin{aligned}\mathbf{D}_{E(2;2)} &= \mathbf{D}_{E(2)\#\overline{\mathbb{C}P}^2} - \mathbf{D}_{E(2)\#\overline{\mathbb{C}P}^2, f-2e}\\ &= \exp(\frac{Q}{2})\cosh(E) - (-1)^{\frac{(f-2e)^2+E\cdot(f-2e)}{2}}\exp(\frac{Q}{2})\cosh(E)\\ &= 2\exp(\frac{Q}{2})\cosh(E).\end{aligned}$$

While E is not a cohomology class in $E(2;2)$, the appropriate class is $E+\frac{1}{2}(F-2E) = \frac{F}{2}$ (since we can add multiples of classes in the neighborhood of the -4 sphere), i.e. the multiple fiber. So

$$\mathbf{D}_{E(2;2)} = 2\exp(\frac{Q}{2})\cosh(\frac{F}{2}) = \exp(\frac{Q}{2})\frac{\sinh(F)}{\sinh(\frac{F}{2})} \tag{9}$$

4.3. The basic computational theorem

To compute $\mathbf{D}_{X_p}$ for large p requires an important preliminary observation about the Donaldson series. For $u \in H_2(X)$ and $F \in \mathbf{A}^*(X)$, the interior product

$$\iota_u F(v) = (\deg(v)+1)F(uv)$$

is a derivation. On the formal power series $\mathbf{F}$ on $H_2(X)$ defined by $\mathbf{F}(\alpha) = F(\exp(\alpha))$ this induces

$$\partial_u \mathbf{F}(\alpha) = F(u\exp(\alpha)).$$

This is just the formal derivative of $\mathbf{F}$ in the direction u. Similarly, for higher order derivatives, $\partial_u^k \mathbf{F}(\alpha) = F(u^k\exp(\alpha))$. Note that the linearity of F implies that $\partial_{u+v}\mathbf{F} = \partial_u\mathbf{F} + \partial_v\mathbf{F}$.

An induction argument shows that

$$\partial_u^{2k}\exp(Q/2) = \exp(Q/2)\sum_{t=0}^{k}(u\cdot u)^{k-t}\tilde{u}^{2t}\binom{2k}{2t}\frac{(2k-2t)!}{2^{k-t}(k-t)!} \tag{10}$$

and
(11)

$$\partial_u^{2k+1}\exp(Q/2) = \exp(Q/2)\sum_{t=0}^{k}(u\cdot u)^{k-t}\tilde{u}^{2t+1}\binom{2k+1}{2t+1}(2k-2t)!2^{k-t}(k-t)!$$

where $\tilde{u}$ is the dual form of $u \in H_2(X)$ with respect to the intersection form Q. For $\kappa \in H_2(X)$ we have

(12) $$D_u(e^\kappa) = (\kappa\cdot x)e^\kappa.$$

The basic computational theorem for rational blowdowns is:

Theorem 4.3.1 ([**FS10**]). *Let X be a simply connected 4-manifold of simple type containing the configuration C_p, and let X_p be the result of rationally blowing down C_p. Then, restricted to $X_c = X_p \setminus B_p = X \setminus C_p$:*

$$\mathbf{D}_{X_p} = \sum_{i=1}^{m(p)}\alpha_i(p)\partial^{n_i(p)}\mathbf{D}_{X,c_i(p)}$$

where $\alpha_i(p) \in \mathbf{Q}$, $\partial^{n_i(p)}$ is an n_ith order partial derivative with respect to classes in $H_2(C_p;\mathbf{Z})$ and $c_i(p) \in H_2(C_p;\mathbf{Z})$, and these quantities depend only on p, not on X.

It is routine to derive Theorem 1.3.5 in the first lecture from this theorem.

4.4. The Donaldson series for $E(n;p,q)$

Recall that for $n \geq 2$:

$$\mathbf{D}_{E(n)} = \exp(Q/2)\sinh^{n-2}(f)$$

where f is the class of a fiber. Let $X = E(2)\#(p-1)\overline{\mathbf{CP}}^2$, and let X_p be the rational blowdown of $C_p \subset X$, so that $X_p \cong E(2;p)$. Since $\mathbf{D}_{E(2)} = \exp(Q/2)$, the blowup formula yields

(13) $$\mathbf{D}_X = \frac{1}{2^{p-1}}\exp(Q/2)\sum_J \exp(\sum_{i=1}^{p-1}\epsilon_{J,i}e_i)$$

where the outer sum is taken over all $J = (\epsilon_{J,1},\ldots,\epsilon_{J,p-1}) \in \{\pm 1\}^{p-1}$. The basic classes of X are $\{\kappa_J = \sum \epsilon_{J,i}e_i\}$, and applying Theorem 1.3.5 we get

(14) $$\mathbf{D}_{X_p} = \exp(\bar{Q}/2)\sum_J b_J e^{\bar{\kappa}_J}$$

where $\bar{\kappa}_J \in H_2(X_p;\mathbf{Z})$ is the unique extension of $\kappa_J|_{X_c}$. Recall that the spheres of the configuration C_p represent homology classes $u_i = e_{p-(i+1)} - e_{p-i}$ for $1 \leq i \leq p-2$, and $u_{p-1} = f - 2e_1 - e_2 - \cdots - e_{p-1}$. In X_p we have the multiple fiber $f_p = f/p$.

Proposition 4.4.1. $\bar{\kappa}_J = |J|\cdot f_p$ *where* $|J| = \sum_{i=1}^{p-1}\epsilon_{J,i}$.

Proof. First we find a class $\zeta \in H_2(C_p;\mathbf{Q})$ so that $(\kappa_J+\zeta)\cdot u_i = 0$ for each i. This means that $\kappa_J+\zeta \in H_2(X_c;\mathbf{Q})$, and as dual forms: $H_2(X_c;\mathbf{Z}) \to \mathbf{Z}$, $\kappa_J|_{X_c} = \kappa_J+\zeta$. To find ζ we need to solve the linear system

$$(\kappa_J + \sum x_i u_i)\cdot u_j = 0, \quad j = 1,\ldots,p-1.$$

We begin by reexpressing these equations. Let $\{\omega_i\}$ be a standard basis for $\mathbf{Q}^{p-1}$, and let A be the $(p-1)\times(p-1)$ matrix whose ith row vector is

$$\begin{aligned} A_i &= \omega_{p-(i+1)} - \omega_{p-i}, \quad i=1,\dots,p-2 \\ A_{p-1} &= -2\omega_1 - \omega_2 - \cdots - \omega_{p-1} \end{aligned}$$

We have $u_i = A^t(\omega_i)\cdot \mathbf{e}$ and $u_{p-1} = f + A^t(\omega_{p-1})\cdot\mathbf{e}$, where $\mathbf{e}=(e_1,\dots,e_{p-1})$. Our linear system is equivalent to

$$P\mathbf{x} = A\boldsymbol{\epsilon}_J$$

where $\mathbf{x}=(x_1,\dots,x_{p-1})$ and $\boldsymbol{\epsilon}_J=(\epsilon_{J,1},\dots,\epsilon_{J,p-1})$. (The matrix P is the plumbing matrix for C_p.) Hence $\mathbf{x}=P^{-1}A\boldsymbol{\epsilon}_J$.

We claim that $P(A^t)^{-1}=-A$. This can be checked on the basis

$$\{\omega_2-\omega_1,\dots,\omega_{p-1}-\omega_{p-2},\omega_{p-1}\}$$

using

$$A(\omega_i) = -\omega_{p-1} - \omega_{p-(i+1)} + \omega_{p-i}, \ 2\le i\le p-1 \ (\omega_0=0),$$

$$A(\omega_1) = -2\omega_{p-1} + \omega_{p-2}$$

$$P(\omega_i) = \omega_{i+1} - 2\omega_i + \omega_{i-1}, \ i\neq p-1$$

$$P(\omega_{p-1}) = -(p+2)\omega_{p-1} + \omega_{p-2}.$$

It follows that $A^tP^{-1}A=-I$. Thus

$$\kappa_J+\zeta = \kappa_J + \sum x_iu_i = (\epsilon_J + A^t\mathbf{x})\cdot\mathbf{e} + x_{p-1}f = (\boldsymbol{\epsilon}_J - \boldsymbol{\epsilon}_J)\cdot\mathbf{e} + x_{p-1}f = x_{p-1}f.$$

To compute x_{p-1} note that

$$A\boldsymbol{\epsilon}_J = (\epsilon_{J,p-2}-\epsilon_{J,p-1}, \epsilon_{J,p-3}-\epsilon_{J,p-2},\dots,\epsilon_{J,1}-\epsilon_{J,2}, -2\epsilon_{J,1}-\epsilon_{J,2}-\cdots-\epsilon_{J,p-1})$$

so that if $(P^{-1})_{p-1}$ denotes the bottom row of P^{-1}:

$$x_{p-1} = (P^{-1})_{p-1}(A\boldsymbol{\epsilon}_J) = -\frac{1}{p^2}(1,2,\dots,p-1)\cdot(A\epsilon_J) = \frac{1}{p}\sum \epsilon_{J,i} = \frac{1}{p}|J|.$$

Thus $\kappa_J|_{X_c} = \kappa_J+\zeta = \frac{1}{p}|J|f$ as forms: $H_2(X_c;\mathbf{Z})\to\mathbf{Z}$. The homology class $\kappa_J+\zeta$ is in fact an integral class $\bar\kappa_J = |J|f_p\in H_2(X_p;\mathbf{Z})$ which is the unique extension of $\kappa_J|_{X_c}$

In an arbitrary smooth 4-manifold X, define a *nodal fiber* to be an immersed 2-sphere S with one singularity, a positive double point, such that the regular neighborhood of S is diffeomorphic to the regular neighborhood of a nodal fiber in an elliptic surface. (There need not be any associated ambient fibration of X.) Given such a nodal fiber S, one can perform a 'log transform' of multiplicity p by blowing up to get $C_p\subset X\#(p-1)\overline{\mathbf{CP}}^2$ with $u_{p-1}=S-2e_1-e_2-\cdots-e_{p-1}$, and then blowing down C_p. We denote the result of this process by X_p.

Throughout, we use the following notation. If X has simple type, and

$$\mathbf{D}_X = \exp(Q/2)\sum a_s e^{\kappa_s},$$

then we write $\mathbf{K}_X=\sum a_s e^{\kappa_s}$.

Proposition 4.4.2. *Let S be a nodal fiber which satisfies $S \cdot \lambda_j = 0$ for each basic class λ_j of X. Then*

$$\mathbf{D}_{X_p} = \begin{cases} \exp(Q_{X_p}/2)\mathbf{K}_X \cdot (b_{p,0} + \sum_{i=1}^{\frac{p-1}{2}} b_{p,2i}(e^{2iS/p} + e^{-2iS/p})), & p \text{ odd} \\ \exp(Q_{X_p}/2)\mathbf{K}_X \cdot (\sum_{i=1}^{\frac{p}{2}} b_{p,2i-1}(e^{(2i-1)S/p} + e^{-(2i-1)S/p})), & p \text{ even} \end{cases}$$

where the coefficients $b_{p,j}$ depend only on p, not on X.

Proof. The Donaldson series of $X\#(p-1)\overline{\mathbf{CP}}^2$ is

$$\frac{1}{2^{p-1}}\mathbf{D}_X \cdot \exp(Q_{(p-1)\overline{\mathbf{CP}}^2}/2) \sum_J \exp(\sum_{i=1}^{p-1} \epsilon_{J,i} e_i).$$

Theorem 4.3.1 states that $\mathbf{D}_{X_p}$ is obtained from this by applying a differential operator which by hypothesis evaluates trivially on $\mathbf{D}_X$. The proposition now follows from (14) and Proposition 4.4.1 by the Leibniz rule. (That the coefficients of e^{mp} and e^{-mp} are equal follows from the fact that $\mathbf{D}_{E(2;p)}$ is an even function.)

From our computation of the order 2 log transform (9), Proposition 4.4.2 now implies:

Corollary 4.4.3. *If S is a nodal fiber in X orthogonal to all basic classes and X_2 is the multiplicity 2 log transform of X formed from S, then*

$$\mathbf{D}_{X_2} = \exp(Q_{X_2}/2)\mathbf{K}_X \cdot (e^{S/2} + e^{-S/2}). \qquad \square$$

Lemma 4.4.4. *Let X contain a nodal fiber S orthogonal to all basic classes. Then the sum of the coefficients $b_{p,j}$ in the expression for $\mathbf{D}_{X_p}$ in Proposition 4.4.2 is equal to p.*

Proof. In Example 3 we showed that there is a configuration $C'_p \subset X\#(p-1)\overline{\mathbf{CP}}^2 = Y$ where $u'_i = e_{p-(i+1)} - e_{p-i}$ for $i = 1, \dots, p-2$, and $u'_{p-1} = -2e_1 - e_2 - \dots - e_{p-1}$ such that the rational blowdown $Y_p = X\#H_p$ where H_p is a homology 4-sphere with $\pi_1 = \mathbf{Z}_p$. It follows easily that $\mathbf{D}_{Y_p} = p \cdot \mathbf{D}_X$.

As above, we let $\kappa_J = \sum \epsilon_{J,i} e_i$, $J \in \{\pm 1\}^{p-1}$; so

$$\mathbf{D}_Y = \frac{1}{2^{p-1}}\mathbf{D}_X \cdot \exp(Q_{(p-1)\overline{\mathbf{CP}}^2}/2) \sum_J e^{\kappa_J}.$$

All partial derivatives of $\mathbf{D}_X$ with respect to classes in $H_2(C'_p)$ are trivial; so

$$p\mathbf{D}_X = \mathbf{D}_{Y_p} = \mathbf{D}_X \cdot \sum_J b_J e^{\bar{\kappa}_J}.$$

The proof of Proposition 4.4.1 shows that each $\bar{\kappa}_J = 0$; so $\sum_J b_J = p$.

We can also form the configuration $C_p \subset Y$ whose blowdown is the p-log transform of the nodal fiber $S \subset X$. The configurations C_p, C'_p agree, $u_i = u'_i$, except that $u_{p-1} = u'_{p-1} + S$. However, since S is orthogonal to all the basic classes of X, for all i, all intersections of u_i and u'_i with all basic classes of $Y = X\#(p-1)\overline{\mathbf{CP}}^2$ agree. Thus, according to Theorem 1.3.5, the coefficients b_J are the same coefficients that arise in the formula

$$\mathbf{D}_{X_p} = \exp(Q_{X_p}/2)\mathbf{K}_X \sum_J c_J e^{|J|S/p}.$$

This means that the sum of the coefficients of the expression for $\mathbf{D}_{X_p}$ in Proposition 4.4.2 is $\sum_J b_J = p$.

An interesting observation is that if p is any positive odd integer, then a multiplicity $2p$ log transform can be obtained as the result of either a multiplicity p log transform on a nodal fiber of multiplicity 2, or by a multiplicity 2 log transform on a nodal fiber of multiplicity p. Thus

$$\begin{aligned}\mathbf{D}_{E(n;2p)} &= \exp(Q/2)(e^{f_2}+e^{-f_2})(b_{p,0}+\sum_{i=1}^{(p-1)/2} b_{p,2i}(e^{2if_2/p}+e^{-2if_2/p}))\\ &= \exp(Q/2)(b_{p,0}+\sum_{i=1}^{(p-1)/2} b_{p,2i}(e^{2if_p}+e^{-2if_p}))(e^{f_p/2}+e^{-f_p/2})\end{aligned}$$

since we already know the formula for a log transform of multiplicity 2. We compare coefficients using $f_2 = pf_{2p}$, $f_p/2 = f_{2p}$, and $f_p = 2f_{2p}$.

Assume for the sake of definiteness that $p \equiv 1 \pmod 4$ and let $r = (p-1)/4$. In the top expansion, the coefficient of $e^{\pm pf_{2p}}$ is $b_{p,0}$ and $b_{p,2j}$ is the coefficient of $e^{\pm(p+2j)f_{2p}}$ and $e^{\pm(p-2j)f_{2p}}$. In the second expansion, the coefficient of $e^{\pm f_{2p}}$ is $b_{p,0}$, and $b_{p,2j}$ is the coefficient of $e^{\pm(4j-1)f_{2p}}$ and $e^{\pm(4j+1)f_{2p}}$. To simplify notation, let $(m)_1$ be the coefficient of $e^{mf_{2p}}$ in the top expansion and $(m)_2$ its coefficient in the bottom expansion. Then,

$$\begin{aligned}b_{p,0} &= (p)_1 = (p)_2 = b_{p,2r} = (p-2)_2 = (p-2)_1 = b_{p,2} = (p+2)_1\\ &= (p+2)_2 = b_{p,2(r+1)} = (p+4)_2 = (p+4)_1 = b_{p,4} = (p-4)_1 = (p-4)_2\\ &= b_{p,2(r-1)} = (p-6)_2 = (p-6)_1 = b_{p,6} = \cdots\end{aligned}$$

and we see inductively that when p is odd, all the $b_{p,2i}$ are equal. Lemma 4.4.4 now implies

$$b_{p,0} + 2\sum_{i=1}^{(p-1)/2} b_{p,2i} = p.$$

It follows that each $b_{p,2i} = 1$, $i = 0, \ldots, (p-1)/2$.

Similarly, if p is even, let $q = p-1$. Expanding $\mathbf{D}_{E(n;pq)}$ we see that all $b_{p,2i-1}$, $i = 1, \ldots, p/2$ are equal; and so again each $b_{p,2i-1} = 1$.

Theorem 4.4.5. *Let X be a 4-manifold of simple type and suppose that X contains a nodal fiber S orthogonal to all its basic classes. Then*

$$\mathbf{D}_{X_p} = \exp(Q_{X_p}/2)K_X \cdot \frac{\sinh(S)}{\sinh(S/p)}.$$

Proof. If, e.g., p is odd,

$$\begin{aligned}\mathbf{D}_{X_p} &= \exp(Q_{X_p}/2)\mathbf{K}_X \cdot (1 + 2\cosh(2S/p) + 2\cosh(4S/p) + \cdots\\ &+ 2\cosh((p-1)S/p))\\ &= \exp(Q_{X_p}/2)\mathbf{K}_X \cdot \frac{\sinh(S)}{\sinh(S/p)}\end{aligned}$$

As a result we have the calculation of the Donaldson series for all simply connected elliptic surfaces with $p_g \geq 1$.

Theorem 4.4.6. *If $n \geq 2$ and $p, q \geq 1$ are relatively prime,*

$$\mathbf{D}_{E(n;p,q)} = \exp(Q/2)\frac{\sinh^n(f)}{\sinh(f_p)\sinh(f_q)}. \qquad \square$$

This formula was originally conjectured by Kronheimer and Mrowka [**KM1**].

LECTURE 5
Taut Configurations and Horikowa Surfaces

5.1. Taut configurations

Consider a 4-manifold X of simple type containing the configuration C_p. By Theorem 1.3.3 for each 2-sphere u_i in C_p and each basic class κ of X, we have

$$-2 \geq u_i^2 + |u_i \cdot \kappa| \tag{15}$$

except in the special case described in Theorem 1.3.4 where $0 \geq u_i^2 + |u_i \cdot \kappa|$. The only examples known where the special case arises are in blowups. This was the situation in the previous section where we studied log transforms. In this section, we assume that we are not in the special case. We say that a configuration is *tautly embedded* if (15) is satisfied for each u_i of the configuration and each basic class κ of X. Thus, if C_p is tautly embedded, then for every basic class κ, $u_i \cdot \kappa = 0$ for $i = 1, \ldots, p-2$ and $|u_{p-1} \cdot \kappa| \leq p$.

Theorem 5.1.1. *Suppose that X is of simple type and contains the tautly embedded configuration C_p. If*

$$\mathbf{D}_X = \exp(Q/2) \sum a_s e^{\kappa_s}$$

then the rational blowdown X_p satisfies

$$\mathbf{D}_{X_p} = \exp(\bar{Q}/2) \sum \bar{a}_s e^{\bar{\kappa}_s}$$

where

$$\bar{a}_s = \begin{cases} 2^{p-1} a_s, & |u_{p-1} \cdot \kappa_s| = p \\ 0, & |u_{p-1} \cdot \kappa_s| < p \end{cases}$$

Furthermore, if $|u_{p-1} \cdot \kappa_s| = p$ then $\bar{\kappa}_s^2 = {\kappa_s}^2 + (p-1)$.

Proof. An algebraic topological observation is that if $\kappa_s \cdot u_{p-1} \neq 0, \pm p$ then $\bar{a}_s = 0$. For $\kappa_s \cdot u_{p-1} = 0$ note that since the κ_s are characteristic p must be even. For the model, consider Example 4; the smooth 4-manifold $Y = E(p+1) \# (p-1)\overline{\mathbf{CP}}^2$ contains the configuration C'_p with $u'_i = e_{p-i} - e_{p-(i+1)}$ for $1 \leq i \leq p-2$ and $u'_{p-1} = s + e_1$ where s a section of $E(p+1)$ (oriented so that $s \cdot f = 1$) which represented by an embedded sphere of square $s^2 = -(p+1)$. Furthermore, the rational blowdown Y_p of Y is diffeomorphic to $E(p+1)$. Since p is even, the basic classes for Y are of the form $rf \pm e_1 \cdots \pm e_{p-1}$ with f the fiber class and $|r| \leq p-1$, $r \equiv 1 \mod 2$. The basic classes for Y_p are just the classes rf. Consider the basis class $\kappa = f + e_1 + \cdots + e_{p-1}$. Then $\kappa \cdot u_i' = 0$ for all i and $\kappa \in H_2(Y_c;\mathbf{Z})$. Thus, using the notation of Theorem 1.3.5, $\bar{\kappa} = \kappa$ so that coefficient b of $\bar{\kappa}$ in the Donaldson series for $Y_p = E(p+1)$ must vanish. Comparing this model situation $C_p' \subset Y$ with our given situation: $C_p \subset X$ and $\kappa_s \cdot u_{p-1} = 0$, $\kappa_s \neq 0$. Similarly, $\kappa_s \in H_2(X_c;\mathbf{Z})$; so Theorem 1.3.5 implies that $\bar{a}_s = c_0 = 0$.

In case $\kappa_s \cdot u_{p-1} = \pm p$, we compare with the model for the order p log transforms of $E(2)$; $C_p'' \subset Y = E(2)\#(p-1)\overline{\mathbf{CP}}^2$ which is blown down to obtain $Y_p = E(2;p)$. Again let $\lambda_0 = \pm(e_1 + \cdots + e_{p-1})$; so by Lemma 4.4.1, λ_0 is the unique basic class of Y_p satisfying $\bar{\lambda}_0 = \pm(p-1)f_p \in H_2(Y_p;\mathbf{Z})$. Now

$$\begin{aligned} \mathbf{D}_Y &= \frac{1}{2^{p-1}}\exp(Q/2)\sum \exp(\pm e_1 \pm \cdots \pm e_{p-1}) = \exp(Q/2)\sum_J \frac{1}{2^{p-1}} e^{\lambda_J} \\ \mathbf{D}_{Y_p} &= \exp(Q/2)\sum{}_{\ell \equiv p \pmod 2}^{|\ell| \leq p-1} e^{\ell f_p} = \exp(Q/2)\sum \frac{1}{2^{p-1}} b_J e^{\lambda_J} \end{aligned}$$

Since λ_0 is the unique λ_J with $\bar{\lambda}_0 = \pm(p-1)f_p$, we get $b_0 = 2^{p-1}$. We may now apply Theorem 1.3.5 to obtain our result since $\kappa_s \cdot u_i = \lambda_0 \cdot u_i''$ for each i.

In order to compute $\bar{\kappa}_s^2$, we find $x_i \in \mathbf{Q}$, $1 = 1, \ldots, p-1$, such that

$$\kappa_s + \zeta = \kappa_s + \sum_{i=1}^{p-1} x_i u_i \in H_2(C_p;\mathbf{Q})$$

as in the proof of Proposition 4.4.1. We can solve for the x_i using the model $C_p'' \subset Y''$, and $\boldsymbol{\epsilon}_J = \pm(1,\ldots,1)$ in the proof of Proposition 4.4.1. Referring there, we get

$$\mathbf{x} = P^{-1}A\boldsymbol{\epsilon}_J = -(A^t)^{-1}\boldsymbol{\epsilon}_J = \pm\frac{1}{p}(1,2,\ldots,p-1).$$

So $\zeta = \pm\sum \frac{i}{p}u_i$, and $\zeta^2 = \mathbf{x}\cdot P\mathbf{x} = 1-p$. Hence

$$\bar{\kappa}_s^2 = (\kappa_s + \zeta)^2 = \kappa_s^2 + 2\kappa_s \cdot \zeta + \zeta^2 = \kappa_s^2 + (p-1).$$

5.2. Horikowa surfaces

To further illustrate the utility of Theorem 5.1.1 we compute the Donaldson invariants of a family of Horikawa surfaces $\{H(n)\}$ with $c_1(H(n))^2 = 2n-6$. To obtain $H(n)$, start with the simply connected ruled surface $\mathbf{F}_{n-3}$ whose negative section s_- has square $-(n-3)$. We have seen in the first hint for Exercise 13 that the classes $s_+ + f$ and s_- form a configuration in $\mathbf{F}_{n-3}$ whose regular neighborhood D_{n-2} has complement the rational ball B_{n-2}. The Horikawa surface $H(n)$ is defined to be the 2-fold branched cover of $\mathbf{F}_{n-3}$ branched over a smoothing of $4(s_+ + f) + 2s_-$.

Lemma 5.2.1. *For $n \geq 4$, the elliptic surface $E(n)$ contains a pair of disjoint configurations C_{n-2} in which the spheres u_{n-1} are sections of $E(n)$ and for $1 \leq j \leq$*

$n-2$, $u_j \cdot f = 0$. *Furthermore, the rational blowdown of this pair of configurations is the Horikawa surface* $H(n)$.

Proof. It follows from our description of $H(n)$ that there is a decomposition

$$H(n) = B_{n-2} \cup \tilde{D}_{n-2} \cup B_{n-2}$$

where $\tilde{D}_{n-2}$ is the branched cover of D_{n-2}. Rationally blow up each B_{n-2}; this is then the 2–fold branched cover of $\mathbf{F}_{n-3}$ with B_{n-2} blown up. The result is the complex surface $C_{n-2} \cup \tilde{D}_{n-2} \cup C_{n-2}$ which, by computing characteristic numbers, is just $E(n)$. The first case $n = 4$ gives the example $H(4) = W_2$ above. The Horikawa surfaces $H(n)$ lie on the Noether line $5c_1^2 - c_2 + 36 = 0$, and of course the elliptic surfaces $E(n)$ lie on the line $c_1^2 = 0$ in the plane of coordinates (c_1^2, c_2). Let $Y(n)$ be the simply connected 4-manifold obtained from $E(n)$ by blowing down just one of the configurations C_{n-2}. Then $c_1(Y(n))^2 = n-3$ and $c_2(Y(n)) = 11n+3$; so $Y(n)$ lies on the bisecting line $11c_1^2 - c_2 + 36 = 0$. The calculation of Donaldson invariants of $Y(n)$ and $H(n)$ follows directly from Theorem 5.1.1.

Proposition 5.2.2. *The Donaldson invariants of* $Y(n)$ *and* $H(n)$ *are:*

$$\mathbf{D}_{Y(n)} = \begin{cases} \exp(Q/2)\sinh(\lambda_n), & n \text{ odd} \\ \exp(Q/2)\cosh(\lambda_n), & n \text{ even} \end{cases}$$

$$\mathbf{D}_{H(n)} = \begin{cases} 2^{n-3}\exp(Q/2)\sinh(\kappa_n), & n \text{ odd} \\ 2^{n-3}\exp(Q/2)\cosh(\kappa_n), & n \text{ even} \end{cases}$$

where $\lambda_n^2 = n-3$ *and* $\kappa_n^2 = 2n-6$. □

Corollary 5.2.3. *The simply connected 4-manifolds* $Y(n)$ *are not homotopy equivalent to any complex surface.*

Proof. If $Y(n)$ were homeomorphic to a complex surface, this computation shows that it would have to be minimal, since the formula for $\mathbf{D}_{Y(n)}$ does not contain a factor $\cosh(e)$ where $e^2 = -1$. Certainly the surface in question could not be elliptic since $c_1(Y(n))^2 \neq 0$. But neither could the surface be of general type since $Y(n)$ violates the Noether inequality. Thus $Y(n)$ is not homeomorphic to any complex surface.

D. Gomprecht [**Gt**] has computed the value of the Donaldson invariant $D_X(F^k)$ for any Horikawa surface X and k large, where F is the branched cover of the fiber f of F_{n-3}.

Certain of the Horikowa surfaces have two deformation types. Let $K(r)$ be the double cover of $\mathbf{F}_0$ branched over a smoothing of $6s_+ + 4rf$ and let $K'(r)$ be the double cover of $\mathbf{F}_{2r}$ branched over a disconnected branch locus which is a smoothing of $5s_+ \coprod s_-$. These are surfaces with $c_1^2 = 8r-8$ and $c_2 = 40r-4$ and Horikowa has shown that they are deformation inequivalent. If r is even these surfaces can be distinguished by the type of their intersection form. However, if r is odd, both these surfaces have odd intersection form and hence are homotopy equivalent.

Problem 20. Are the two deformation types of Horikowa surfaces diffeomorphic?

A related problem is

Problem 21. Are there restrictions on self-diffeomorphisms f of a minimal Kähler surface X (of nonnegative Kodaira dimension) with canonical class K_X beyond the conditions $f^*K_X = \pm K_X$? For example, consider the surface which is the double cover of $S^2 \times S^2$ along a branch locus which represents $|2af_1 + 2bf_2|$, where the f_i are the fibers of the two different projections and a, b are positive integers with $a \neq b$. Is it true that for general a and b every orientation-preserving self-diffeomorphism of X preserves the pullbacks to X of f_1 and f_2 up to sign and not just K_X (which is a positive combination of the pullbacks)? Even for simply connected elliptic surfaces there is a gap of finite index between those isometrics of $H^2(X; \mathbf{Z})$ known to arise from self-diffeomorphisms and the restrictions placed on such isometrics by Donaldson theory or Seiberg-Witten theory (namely, that they preserve the homology orientation of the manifold). So the problem is to construct more diffeomorphisms for these double branched covers or to construct new invariants which rule out the existence of such diffeomorphisms.

Recall that $E(n)$ has nine disjoint sections (with self-intersection $-n$). It would be interesting if we could complete these to nine disjoint configurations $C(n-2)$. Let P_n be the result of blowing down all nine configurations.

Exercise 35. If $n > 27$, then P_n has positive signature.

Exercise 36. Show $E(5)$ and $E(6)$ have seven such configurations and compute the Donaldson series of the 4-manifold obtained by blowing down these configurations. Generalize your construction.

Problem 22. Determine if all simply-connected smooth 4-manifolds are obtained from the $E(n)$ by a sequence of blowups, rational blowups, blowdowns, and rational blowdowns.

Exercise 37. Let X be a simply-connected smooth 4-manifold with $b^+ > 1$ and with simple-type. Let $\mathcal{B}$ denote the collection of homology classes $K \in H_2(X; \mathbf{Z})$ which satisfy the adjunction inequality $2k - 2 \geq x^2 + |x \cdot K|$ for every $x \in H_2(X; \mathbf{Z})$. Here k can be either the number of positive double points of an immersed 2-sphere representing x or the genus of a surface representing x (you choose). Show that $\mathcal{B}$ is finite and is a diffeomorphism invariant of X.

Problem 23. Relate $\mathcal{B}$ with the set of basic classes for X.

Problem 24. Does there exist a homology 3-sphere Σ which bounds a smooth spin 4-manifold W with $\sigma(W) \equiv 8 \mod 16$ and such that $\Sigma \# \Sigma$ bounds an acyclic smooth 4-manifold?

References

[APS1] M. Atiyah, V. Patodi, and I. Singer, *Spectral asymmetry and Riemannian geometry. I*, Math. Proc. Camb. Phil. Soc. **77** (1975), 43–69.

[Ak] N. Akhiezer, "Elements of the Theory of Elliptic Functions," translated by H. McFaden, A.M.S. Translations of Math. Monographs 79 (1990).

[Au] D. Austin, *$SO(3)$-instantons on $L(p,q) \times \mathbb{R}$*, Jour. Diff. Geom. **32** (1990), 383–413.

[AB] D. Austin and P. Braam, *Equivariant Floer (co)homology and gluing Donaldson polynomials*, preprint.

[B] S. Bauer, *Diffeomorphism types of elliptic surfaces with $p_g = 1$*, J. Reine Angew. Math. **451** (1992), 149–180.

[Br] E. Brieskorn, *Bespiele zur Differentialtopologie von singularitaten*, Inv. Math. **2** (1966), 1–14.

[BPV] W. Barth, C. Peters, and A. Van de Ven, "Compact Complex Surfaces," Springer-Verlag, 1984.

[B] J. Bryan, *Symplectic geometry and the relative Donaldson invariants of $\overline{\mathbf{CP}}^2$*, 1994, Harvard University Thesis.

[C] Z. Chen, *On the geography of surfaces*, Math. Annalen 277 (1987), 141–164.

[Dv] I. Dolgachev *Weighted projective varieties*, in 'Group Actions and Vector Fields,' Springer Lecture Notes No.956, 1982, pp.34–71.

[D1] S. Donaldson, *An application of gauge theory to the topology of 4-manifolds*, Jour. Diff. Geom. **18** (1983), 269–316.

[D2] S. Donaldson, *Anti-self-dual Yang-Mills connections on complex algebraic surfaces and stable vector bundles*, Proc. Lon. Math. Soc. **3** (1985), 1–26.

[D3] S. Donaldson, *Connections, cohomology and the intersection forms of 4-manifolds*, Jour. Diff. Geom. **24** (1986), 275–341.

[D4] S. Donaldson, *Irrationality and the h-cobordism conjecture*, Jour. Diff. Geom. **26** (1987), 141–168.

[D5] S. Donaldson, *The orientation of Yang-Mills moduli spaces and 4-dimenhsional topology*, Jour. Diff. Geom. **26**(1987), 397–428.

[D6] S. Donaldson, *Polynomial invariants for smooth 4-manifolds*, Topology **29** (1990), 257–315.

[D7] S. Donaldson, *Yang-Mills invariants of 4-manifolds*, Geometry of Low-Dimensional Manifolds I, London Math. Society Lecture Note Ser. **150** (1989), Cambridge University Press, 257–315.

[DK] S. Donaldson and P. Kronheimer, "The Geometry of Four-Manifolds," Oxford Mathematical Monographs, Oxford University Press, Oxford, 1990.

[E2] W. Ebeling, *An example of two homeomorphic, nondiffeomorphic complete intersection surfaces*, Inv. Math. **99** (1990), 651–654.

[EO1] W. Ebeling and C. Okonek, *Donaldson invariants, monodromy groups, and singularities*, Int. Jour. Math. **1** (1990), 233–250.

[FS1] R. Fintushel and R. Stern, *$SO(3)$ connections and the topology of 4-manifolds*, J. Diff. Geom.**20** (1984) 523–539.

[FS2] R. Fintushel and R. Stern, *Definite 4-manifolds*, J. Diff. Geom.**28** (1988) 133–141.

[FS3] R. Fintushel and R. Stern, *Instanton homology of Seifert fibered homology three spheres*, Proc. London Math. Soc. **61** (1990), 109–137.

[FS4] R. Fintushel and R. Stern, *Homotopy $K3$-surfaces containing $\Sigma(2,3,7)$*, J. Diff. Geom.**34** (1991) 255–265.

[FS5] R. Fintushel and R. Stern, *2-Torsion instanton invariants*, Jour. Amer. Math. Soc. **6** (1993), 299–339.

[FS6] R. Fintushel and R. Stern, *Surgery in cusp neighborhoods and the geography of irreducible 4-manifolds*, Inventiones Math., **117** (1994), 455–523.

[FS7] R. Fintushel and R. Stern, *Using the Floer exact triangle to compute Donaldson invariants*, The Floer Memorial Volume, 435–444, Progr. Math. 133, Birkhäuser, Basel 1995.

[FS8] R. Fintushel and R. Stern, *The blowup formula for Donaldson invariants*, Annals of Math. 143 (1996), 529–546.

[FS9] R. Fintushel and R. Stern, *Donalson invariants of 4-manifolds with simple type*, J. Diff. Geom. 42 (1995), 577–633.

[FS10] R. Fintushel and R. Stern, *Rational blow downs of smooth 4-Manifolds* , to appear in J. Diff. Geom.

[F] M. Freedman, *The topology of four-dimensional manifolds*, Jour. Diff. Geom. **17** (1982), 357–454.

[Fr] R. Friedman, *Vector bundles and $SO(3)$-invariants for elliptic surfaces* , Jour. Amer. Math. Soc. **8** (1995), 29–139.

[FM1] R. Friedman and J. Morgan, *On the diffeomorphism types of certain algebraic surfaces I,II*, Jour. Diff. Geom. **27** (1988), 297–369.

[FM2] R. Friedman and J. Morgan, "Smooth Four-Manifolds and Complex Surfaces," Ergebnisse der Math. und Grenz. 3, Springer-Verlag, New York, 1994.

[FMM] R. Friedman, J. Morgan, and B. Moishezon, *On the C^∞ invariance of the canonical class of certain algebraic surfaces*, Bull. Amer. Math. Soc. **17** (1987), 283–286.

[FH] M. Furuta and Y. Hashimoto, *Invariant instantons on S^4*, J. Fac. Sci. Univ. Tokyo **37** (1990), 585–600.

[G1] R. Gompf, *Nuclei of elliptic surfaces*, Topology **30** (1991), 479–511.

[G2] R. Gompf, *Sums of elliptic surfaces*, Jour. Diff. Geom. **34** (1991), 93–114.

[G3] R. Gompf, *A new construction of symplectic manifolds*, Ann. of Math. 142 (1995), 527–595.

[GM] R. Gompf and T. Mrowka, *Irreducible four manifolds need not be complex*, Annals of Math. **138** (1993), 61–111.

[Gt] D. Gomprecht, preprint.

[HZ] F. Hirzebruch and D. Zagier, *The Atiyah-Singer Theorem and Elementary Number Theory*, Mathematics Lecture Series, Vol. 3 (1974), Publish or Perish, Inc.

[Ko] D. Kotschick, *On manifolds homeomorphic to* $\mathbf{CP}^2\#8\bar{\mathbf{CP}}^2$, Inv. Math. **95** (1989), 591–600.

[Ko] D. Kotschick, *$SO(3)$-invariants for 4-manifolds with* $b_2^+ = 1$, Proc. London Math. Soc. **63** (1991), 526–448.

[Kr] P. Kronheimer, *Instanton invariants and flat connections on the Kummer surface*, Duke Math. J. **64** (1991), 229–241.

[KM1] P. Kronheimer and T. Mrowka, *Recurrence relations and asymptotics for four-manifold invariants*, Bull. Amer. Math. Soc. **30** (1994), 215–221.

[KM2] P. Kronheimer and T. Mrowka, *Embedded surfaces and the structure of Donaldson's polynomial invariants*, Jour. Diff. Geom. 41 (1995), 573–734.

[L] T. Leness, *Blowup formulae for $SO(3)$ Donaldson polynomials*, preprint.

[Li] P. Lisca , *On Donaldson polynomials of elliptic surfaces*, Math. Annalen 299 (1994), 629–639.

[Mn1] B. Moishezon, "Complex Surfaces and Connected Sums of Projective Planes," Springer Notes No. 603, 1977.

[Mn2] B. Moishezon, *Analogues of Lefschetz theorems for linear systems with isolated singularities*, Jour. Diff. Geom. **31** (1990), 47–72.

[MM] J. Morgan and T. Mrowka, *A note on Donaldson's polynomial invariants*, Int. Math. Research Notices 10 (1992), 223–230.

[MM2] J. Morgan and T. Mrowka *On the diffeomorphism classification of regular elliptic surfaces*, Int. Math. Research International Math. Research Notices **6** (1993) 183–184.

[MMR] J. Morgan, T. Mrowka, and D. Ruberman, *The L^2-moduli space and a vanishing theorem for Donaldson polynomial invariants*, International Press, 1994.

[MO] J. Morgan and K.G. O'Grady *Differential Topology of Complex surfaces: Elliptic surfaces with $p_g = 1$:Smooth Classification*, Lecture Notes in Mathematics, Vol. 1545, Springer-Verlag, 1993.

[Mr] T. Mrowka, *A local Mayer-Vietoris principle for Yang-Mills moduli spaces*, Ph.D. Thesis, Berkeley, 1988.

[P] U. Persson, *An introduction to the geography of surfaces of general type*, in 'Algebraic Geometry Bowdoin 1985', Proc. Symp. Pure Math. Vol. 46, 1987, pp.195–220.

[R] V.A. Rohlin, *New results in the theory of four dimensional manifolds*, Dok. Akad. Nauk. USSR **84** (1952) 221–224.

[Sa] M. Salvetti, *On the number of nonequivalent differentiable structures on 4-manifolds*, Manuscripta Math. **63** (1989), 157–171.

[SS] A. Stipsicz and Z. Szabo *The smooth classification of elliptic surfaces with* $b^+ > 1$, Duke J. Math.75 (1994), 1–50.

[T1] C. Taubes, "L^2 Moduli Spaces on 4-Manifolds with Cylindrical Ends," International Press, Hong Kong, 1993.

[T2] C. Taubes, *A Mayer-Vietoris model for Donaldson - Floer theory*, Floer Memorial Volume, 586–608, Progr. Math. 133, Birkhauser, Basel 1995.

[T3] C. Taubes, *The role of reducibles in Donaldson - Floer theory*, Proc. 1993 Taniguchi Symposium on Low Dimensional Topology and Topological Field Theory, 171–191.

[T4] C. Taubes, *Donaldson - Floer theory for nontrivial circle bundles*, in preparation.

[U] K. Uhlenbeck, *Connections with L^p bounds on curvature*, Commun. Math. Phys. **83** (1982), 31–42.

[W1] C.T.C. Wall, *On the orthogonal groups of unimodular quadratic forms*, Math. Ann. **147** (1962), 328–338.

[W2] C.T.C. Wall, *Diffeomorphisms of* 4*-manifolds*, Jour. London Math. Soc. **39** (1964) 131–40.

[W3] C.T.C. Wall, *On simply connected* 4*-manifolds* Jour. London Math. Soc. **39** (1964) 141–49.

[Wh] J.H.C. Whitehead *On simply connected 4-dimensional polyhedra*, Comm. Math. Helvetici **45** (1949)48–92.

[Wk] W. Wieczorek, *Donaldson invariants of* 4*-manifolds containing embedded spheres*, Michigan State University Thesis, 1995.

Donaldson-Floer Theory

Clifford H. Taubes
James A. Bryan

IAS/Park City Mathematics Series
Volume 4, 1998

Donaldson-Floer Theory

C. H. Taubes, J. Bryan

Abstract. In these lectures we outline a general approach to the Donaldson-Floer theory of 4-manifolds. The approach is based on a differential form formulation of Donaldson invariants and a fibered product description of the instanton moduli spaces. We show how to define "relative Donaldson invariants" for a large class of 4-manifolds with boundary and show that they take values in the "Floer (co)homology" of the boundary.

[1]Department of Mathematics, Harvard University, Cambridge, Massachusetts 02138.
E-mail address: jbryan@math.uci.edu.
[2]Department of Mathematics, University of California, Irvine, California 92717.
[3]Both authors were supported in part by the mathematical division of the NSF.
[4]1991 *Mathematics Subject Classification.* Primary 57R57, 57N13, 57R55. Secondary 58D29.

LECTURE 1
Introduction

1.1. Motivation

These notes are based on the lectures given by the first author at the IAS/Park City Mathematics Institute during July of 1994. In this and the subsequent lectures we outline the approach to Donaldson-Floer detailed in [**Ta1**], [**Ta2**], and [**Ta3**]. Donaldson-Floer theory refers to the relationship between the Donaldson invariants of a compact, connected oriented 4-manifold X^4 and the Floer cohomology of a smoothly embedded 3-manifold M^3 that splits X:

$$X = X_+ \coprod_{M^3} X_-. \tag{1.1}$$

By this we mean that X is obtained from the two manifolds with boundaries X_+ and X_- by identifying $\partial X_+ = M^3$ with $\partial X_- = -M^3$ ($-M^3$ denotes M^3 with the opposite orientation).

The case where M^3 is a homology sphere is well understood (see [**Fl**] and [**At**]. The approach outlined in these notes is applicable in a much broader setting, in particular when M^3 is a non-trivial circle bundle over a Riemann surface. Splittings by such 3-manifolds arise, for example, as the links of embedded surfaces with non-trivial self-intersection.

Donaldson-Floer theory is motivated by attempting to compute the Donaldson invariants of $X = X_+ \coprod_{M^3} X_-$ in terms of invariants defined solely on each piece.

For an analogy, consider computing the fundamental group of X. By the Seifert-Van Kampen theorem

$$\pi_1(X) \cong \pi_1(X_+) * \pi_1(X_-)/N$$

where $*$ denotes free product and N is the normal subgroup generated by the images of $\pi_1(M)$ in $\pi_1(X_+)$ and $\pi_1(X_-)$. Thus, to compute the invariant $\pi_1(X)$ we combine the invariants from X_+ and X_- using a "gluing rule" coming from M^3.

A closer analogy would be the invariant $R_X \equiv \mathrm{Hom}(\pi_1(X), \mathrm{SO}(3))$, *i.e.* representations of the fundamental group into SO(3). A representation $\rho_X \in R_X$ is uniquely determined by two representations $\rho_{X_+} \in R_{X_+}$ and $\rho_{X_-} \in R_{X_-}$ that agree when restricted to M^3. We thus can write

$$R_X = \{(\rho_{X_+}, \rho_{X_-}) \in R_{X_+} \times R_{X_-} : i_+^*(\rho_{X_+}) = i_-^*(\rho_{X_-})\}$$

where $i_\pm : M^3 \to X_\pm$ are the inclusion maps. This construction is an example of a fiber product.

Definition 1.1.1. Let X, Y, and Z be manifolds with maps $f : X \to Z$ and $g : Y \to Z$. We define the *fiber product* to be

$$X \times_Z Y = \{(x, y) \in X \times Y : f(x) = g(y)\}.$$

The inclusion of $X \times_Z Y$ in $X \times Y$ induces projection maps π_X and π_Y which fit into the following diagram:

$$\begin{array}{ccc} X \times_Z Y & \xrightarrow{\pi_Y} & Y \\ {\scriptstyle \pi_X}\downarrow & & \downarrow{\scriptstyle g} \\ X & \xrightarrow{f} & Z \end{array}$$

Exercise 1.1.2. Suppose that f is transverse to g. Show that $X \times_Z Y$ is a manifold.

The following goal is the primary motivation for Donaldson-Floer theory.

Goal 1.1.3. Given a four manifold X with well-defined Donaldson invariants D_X and a decomposition $X = X_+ \coprod_{M^3} X_-$,define a notion of "relative Donaldson invariants" D_{X_+} and D_{X_-} that take values in well-defined cohomology theories $HF^*(M^3)$ and $HF_*(M^3)$ such that

$$D_X = \langle D_{X_+}, D_{X_-} \rangle$$

where $\langle , \rangle : HF^* \otimes HF_* \to \mathbb{R}$ is the natural cohomology pairing.

In the rest of this lecture we will discuss why there should exist such relative invariants and we will outline the strategy for defining them.

1.2. The moduli space and the invariants

Let X be a simply connected, oriented, compact 4-manifold with $b_2^+(X) > 1$ and odd. For the sake of simplicity of exposition, we will restrict our attention to the SU(2) invariants and we will not use the four dimensional class. In this case the Donaldson invariants are linear maps

$$D_X^d : \mathrm{Sym}^d(H_2(X)) \to \mathbb{R}$$

given by

$$D_X^d(\Sigma^d) = \int_{\mathcal{M}_X^{2d}} \omega_\Sigma^d. \tag{1.2}$$

Here $\mathcal{M}_X^{2d}$ denotes the moduli space of anti-self-dual (ASD) connections (modulo gauge equivalence) on an SU(2) bundle with c_2 such that the dimension of the moduli space is $2d$ (we take the moduli space to be empty if no such bundle exists). ω_Σ is a differential form on $\mathcal{M}_X^{2d}$ of degree 2 that is canonically associated to a representative for the homology classes $[\Sigma] \in H_2(X)$. This form is a representative for the image of Donaldson's μ-map in the sense that ω_Σ is supported in a neighborhood of Donaldson's divisor representation of $\mu(\Sigma)$ and is the Thom class of the normal bundle of the divisor. See [**Ta3**] for the detailed definition of ω_Σ and its properties. The most important property of the form representatives is the following:

Theorem 1.2.1. *The definition of D_X given by integration of the forms ω_Σ as in equation 1.2 coincides with the usual definition*[1].

Proof. If one chooses d surfaces Σ_i representing the class Σ that meet pairwise transversely and integrate the wedge product of the corresponding forms: $\omega_{\Sigma_1} \wedge \cdots \wedge \omega_{\Sigma_d}$, the fact that you will get the usual Donaldson invariant follows simply from the fact that the forms are supported in a neighborhood of the divisor representation and are the Thom class of the normal bundle. The fact that the theorem holds for the d-th wedge product of a single representative ω_Σ requires a detailed understanding of the way the forms behave near the boundary of $\mathcal{M}_X$ (see [**Ta4**]).

Consider now the decomposition 1.1. If we fix a base point $x_0 \in M^3 \subset X$ we have the based moduli space

$$\mathcal{M}_X^0 = \frac{\{\text{ASD connections on } P \to X\}}{\{\text{Gauge transformations of } P \text{ that restrict to the identity on } P_{x_0}\}}.$$

For a generic metric on X, the projection $\mathcal{M}_X^0 \xrightarrow{\pi} \mathcal{M}_X$ is a principal SO(3) fibration and so there exists a 3-form α on $\mathcal{M}_X^0$ with the following properties:

1. α is SO(3) invariant;
2. α is the volume form restricted to each fiber;
3. for any $\theta \in \Omega^{\text{top}}(\mathcal{M}_X)$ we have

$$\int_{\mathcal{M}_X} \theta = \int_{\mathcal{M}_X^0} \alpha \wedge \pi^* \theta.$$

Exercise 1.2.2. Let $P \xrightarrow{\pi} B$ be a principal G fibration with G a compact Lie group of dimension n. Prove that there exists a G invariant form $\alpha \in \Omega^n(P)$ so that α restricted to any fiber is the volume form. Construct an α explicitly from a connection on P. Can α in general be chosen so that $d\alpha = 0$?

The advantage of working with the based moduli space is that it affords a description as a fibered product. Let $\mathcal{B}_M^0$ be the space of connections on $P|_{M^3}$ modulo gauge equivalences based at $x_0 \in M^3 \subset X$. One can make a precise definition of

$$\mathcal{M}_\pm^0 = \frac{\{\text{ASD connections on } X_\pm \text{ that are ``smooth up to the boundary''}\}}{\{\text{Gauge transformations based at } x_0\}}$$

The inclusions $i_\pm : M^3 \to X_\pm$ induce maps

$$i_\pm^* : \mathcal{M}_\pm^0 \to \mathcal{B}_M^0,$$

and also we have restriction maps

$$r_\pm : \mathcal{M}_X^0 \to \mathcal{M}_\pm^0.$$

[1] As with the usual definition, one may have to pass to large enough degree for the invariant to be well-defined and the usual tricks for extending the definition to all degrees apply (see [**Fr-Mo**]).

$\mathcal{M}^0_\pm$ and $\mathcal{B}^0_M$ can be given the structure of infinite dimensional Hilbert manifolds. Assuming M^3 is connected, we have the following

Theorem 1.2.3. $\mathcal{M}^0_X = \mathcal{M}^0_+ \times_{\mathcal{B}^0_M} \mathcal{M}^0_-$.

The theorem tells us that in order to obtain (based gauge equivalence classes of) ASD connections on X from ASD connections on X_+ and X_-, we only need to know that their values agree (up to gauge equivalence) on M^3. In particular, we don't need to require agreement in the derivatives. Fundamentally, the theorem reflects the fact that the anti-self-duality equations are first order.

Remark 1.2.4. If SO(3) acts freely on $\mathcal{B}^0_{M^3}$, *i.e.* $P|_{M^3}$ has no reducible connections, then theorem 1.2.3 will hold for the unbased spaces:

$$\mathcal{M}_X = \mathcal{M}_+ \times_{\mathcal{B}_M} \mathcal{M}_-. \tag{1.3}$$

However, $\mathcal{B}^0_M$ always contains reducibles (the trivial connection for example); even if we include the case where P is an SO(3) bundle (and so possible non-trivial on M) $\mathcal{B}^0_M$ will still contain reducibles. But eventually we will consider metrics on X which will allow us to restrict our attention to neighborhoods of flat connections in $\mathcal{B}^0_M$. Thereby reducibles will be avoided in certain good cases and 1.3 will hold.

1.3. A strategy to compute the invariants

Roughly speaking,the general strategy for computing the integral

$$\int_{\mathcal{M}_X} \omega^d_\Sigma = \int_{\mathcal{M}^0_X} \omega^d_\Sigma \wedge \alpha$$

will be to integrate along the $\mathcal{M}^0_+$ coordinates first and the $\mathcal{M}^0_-$ coordinates second. The added complication of having the form α in the integrand is essentially algebraic in nature and is discussed in [**Ta2**]. For the purposes of these notes we will pretend we are the case where 1.3 holds. Also for simplicity let us only consider D_X on classes $\Sigma \in H_2(X)$ such that $\Sigma = \Sigma_+ + \Sigma_-$ with $\Sigma_\pm \in H_2(X_\pm)$, *i.e.* $\Sigma|_{M^3}$ is 0. Then the form representatives $\omega_{\Sigma_\pm}$ have the property that they are pulled back by $r_\pm$ from forms on $\mathcal{M}_\pm$. Thus the invariant

$$D_X(\Sigma^d) = \int_{\mathcal{M}_X} \omega^d_\Sigma = \int_{\mathcal{M}_+ \times_{\mathcal{B}} \mathcal{M}_-} (\omega_{\Sigma_+} + \omega_{\Sigma_-})^d$$

can be written as a sum of integrals of the form

$$\int_{\mathcal{M}_+ \times_{\mathcal{B}} \mathcal{M}_-} r^*_+\omega_+ \wedge r^*_-\omega_-$$

where $\omega_\pm \in \Omega^*(\mathcal{M}_\pm)$.

To understand what it means to "integrate along the $\mathcal{M}^0_+$ coordinates first and the $\mathcal{M}^0_+$ second" let us consider the finite dimensional, compact model.

Let $\{X, Y, Z, f, g\}$ be as in definition 1.1.1 and assume

- X, Y, and Z are compact, finite dimensional manifolds,
- f is a fibration
- and g is transversal of f.

For fibrations in general, there is a well-defined push-forward map on differential forms:

$$f_* : \Omega^i(X) \to \Omega^{i-(\dim X - \dim Z)}(Z).$$

Its value on $i - (\dim X - \dim Z) = i - d$ vectors at a point $z \in Z$ is given by the prescription:

$$f_*(\omega)_z(v_1, \dots, v_{i-d}) = \int_{f^{-1}(z)} i_{\tilde{v}_1} \cdots i_{\tilde{v}_{i-d}} \omega$$

where $\tilde{v}_i$ denotes a lifting of v_i and $i_{\tilde{v}_i}$ denotes contraction with $\tilde{v}_i$.

Exercise 1.3.1. Show the above construction is well-defined.

We then have

Claim 1.3.2.

$$\begin{aligned} \int_{X \times_Z Y} \pi_X^*(\omega_X) \wedge \pi_Y^*(\omega_Y) &= \int_Y \pi_{Y*} \pi_X^* \omega_X \wedge \omega_Y \\ &= \int_Y g^* f_* \omega_X \wedge \omega_Y. \end{aligned}$$

Proof. Exercise.

The strategy then is clear: to compute the Donaldson invariant we should try to make sense of the equation

$$\int_{\mathcal{M}_+ \times_{\mathcal{B}} \mathcal{M}_-} r_+^* \omega_+ \wedge r_-^* \omega_- = \int_{\mathcal{M}_-} i_-^* (i_+)_* \omega_+ \wedge \omega_-.$$

The relative invariant for X_+ and X_- will then be $(i_+)_*\omega_+$ and $\int_{\mathcal{M}_-} i_-^*(\cdot) \wedge \omega_-$ respectively provided we can

1. make sense of the expressions, and
2. find cohomology theories that are the receptacles for the invariants.

The new features that arise and the technical details to overcome are the main subjects of the subsequent lectures.

LECTURE 2
Quantization

2.1. Problems

Recall from the previous lecture that we can write the based instanton moduli spaces as a fibered product

$$\mathcal{M}_X^0 = \mathcal{M}_+^0 \times_{\mathcal{B}^0} \mathcal{M}_-^0$$

and that to compute the Donaldson invariants of X evaluated on classes coming from classes on X_+ and X_- we would like to make sense of the integral

$$\int_{\mathcal{M}_-^0} i_-^*(i_+)_*\omega_+ \wedge \omega_- . \tag{2.1}$$

First let us see to what extent claim 1.3.2 generalizes. The basic theorem is explained in detail in [**Ta3**] chapter 1, sections e) through o).

Theorem 2.1.1. *Let X, Y, and Z be (possibly infinite dimensional) manifolds with smooth maps $f : X \to Z$ and $g : Y \to Z$ and suppose that f is transverse to g and that f is Fredholm,* i.e.

1. *the differential, df, is a section of the fiber bundle of Fredholm homomorphisms from TX to f^*TZ, and*
2. *the index, $Index(df)$, is oriented in the real K-theory of X, $K_{\mathbb{R}}(X)$.*

Then there is a subcomplex of differential forms on X and a suitable definition of distribution-valued differential forms on Z such that the push-forward map f_* is well-defined. Furthermore, if Y is finite dimensional, then one can make sense of the formula

$$\begin{aligned} \int_{X \times_Z Y} \pi_X^*(\omega_X) \wedge \pi_Y^*(\omega_Y) &= \int_Y \pi_{Y*}\pi_X^*\omega_X \wedge \omega_Y \\ &= \int_Y g^* f_*\omega_X \wedge \omega_Y . \end{aligned}$$

A few remarks:

- If f is proper and $\operatorname{Coker} df_x$ is always 0, then f is a fibration with finite dimensional, compact fibers. The push forward can then be defined again by integration over the fibers and the generalization of claim 1.3.2 to the infinite dimensional setting is straight forward.
- If f fails to be a fibration, *i.e.* $\operatorname{Coker} df_x$ jumps, then integration over the fiber no longer gives a form on Z but it does make sense as a distribution-valued form.
- If f is not proper then the fibers are not compact, so in order to make sense of integration over the fibers, one has to restrict to a certain subcomplex of forms on X. In the gauge theory application, the non-compactness of the fibers arises from the famous bubbling-off phenomenon.

Here is a rather trivial example to illustrate the most basic kind of distribution-valued form that can arise.

Example 2.1.2. Let X, Z, and Y be a point, the interval $[-1, 1]$, and a subinterval $I \subset [-1, 1]$ respectively. Let f map the point to 0 and let g be the inclusion map. The index of df is -1 so f_* should *raise* degree:

$$f_*(1) = \delta_0 dt$$

where δ_0 is the Dirac delta function (distribution). This follows since we want

$$\int_I f_* 1 = \begin{cases} 0 & \text{if } \{0\} \in \text{int} I \\ 1 & \text{if } \{0\} \notin \text{int} I \end{cases}$$

(If $0 \in \partial I$ then g is not transverse to f.)

In order to fit equation 2.1 into the situation of theorem 2.1.1 there are serious problems to be addressed. Coming to terms with the problems will lead to the novel features of the Floer homology. The two main problems are

- The map $i_+ : \mathcal{M}_+^0 \to \mathcal{B}^0$ is an embedding (away from the reducibles) of infinite codimension. The differential di is thus *not* Fredholm.
- $\mathcal{M}_-^0$ is infinite dimensional.

To surmount these problems we will consider special metrics on X that will allow us to restrict the fiber product to $\mathcal{N}^0 \subset \mathcal{B}^0$, a neighborhood of the flat connections. In the components of this neighborhood we will make certain choices of a projections to "half-infinite dimensional subspaces". The map i_+ composed with the projections will then be Fredholm. The cost of restricting to a neighborhood of the flat connections is that the fiber product description of $\mathcal{M}^0$ is no longer true globally, but only on open sets which then cover $\mathcal{M}^0$. Piecing together the push forward and pull backs on the various open sets leads to a more complicated algebraic structure out of which the Floer differential arises. To construct the neighborhood in $\mathcal{B}^0$ and the cover of $\mathcal{M}^0$ we will need a detailed discussion of the way an ASD connection can distribute its curvature.

2.2. Quantization and the Chern-Simons functional

Recall that the Yang-Mills energy of an ASD connection is quantized. For an ASD connection A we have

$$\begin{aligned}\mathcal{E}(A) &= \int_X |F_A|^2 d\mathrm{Vol} \\ &= \int_X \mathrm{tr}(F_A^2) d\mathrm{Vol} \\ &= 8\pi^2 c_2(P)[X] \in 8\pi^2\mathbb{N}.\end{aligned}$$

This "quantization" is independent of the metric on X. For metrics on X containing a long "tube", there will be an important further quantization of energy.

Given a fixed metric g on X and any $L \in \mathbb{R}^+$ we define a new metric g_L that is isometric to $X_+ \amalg M^3 \times [-L, L] \amalg X_-$. We can assume that the original metric g had a neighborhood of M^3 isometric to a product; such g and the associated metrics g_L will still be generic for generic choices of g away from M^3.

For very large L, $|F_A|$ will be very small over most of $[-L, L] \times M^3$. For example, fix δ and let

$$C_\delta = \{t \in [-L, L] \text{ such that } \int_{[t-\frac{1}{2}, t+\frac{1}{2}]\times M^3} |F_A|^2 \geq \delta\}.$$

C_δ is a union of intervals of total length less than $\mathcal{E}/\delta$.

When the curvature of an ASD connection is small, the equation $F_A = dA + A \wedge A$ along with the ellipticity of the ASD equations give control over all the derivatives of F_A:

Theorem 2.2.1 (Uhlenbeck). Given any $\epsilon > 0$, there exist a $\delta > 0$ such that the following is true. Let A be an ASD connection on $[-\frac{1}{2}, \frac{1}{2}] \times M^3$ with

$$\int_{[-\frac{1}{2}, \frac{1}{2}]\times M^3} |F_A|^2 < \delta.$$

Then there exists a flat connection Γ, a gauge transformation g, and a 1-form a such that $g^*(A) = \Gamma + a$ on $[-1/4, 1/4] \times M^3$ and $||a||_{C^m} < \epsilon$, for all m.

For a fixed bundle P, we can find an L such that for any ASD connection A there are strips of width 1 where the curvature is small. Suppose that we could find a fixed $t_0 \in [-L, L]$ such that $|F_A|$ was small in a strip of width 1 about t_0 for every ASD connection. Then we would be able to decompose

$$\mathcal{M}_X^0 = \coprod_{\sigma \in \pi_0(\rho_{M^3})} \mathcal{M}_{X,\sigma}^0$$

where $\pi_0(\rho_{M^3})$ is the set of components of the flat connections on M^3 and $\mathcal{M}_{X,\sigma}^0$ has a decomposition

$$\mathcal{M}_{X,\sigma}^0 = \mathcal{M}_{+,\sigma}^0 \times_{\mathcal{N}_\sigma^0} \mathcal{M}_{-,\sigma}^0.$$

$\mathcal{N}_\sigma^0 \subset \mathcal{B}^0$ is a small neighborhood of σ. Uhlenbeck's theorem would allow us to take a *regular* neighborhood. For example, if σ consists of a single point Γ, we could choose a diffeomorphism

$$\mathcal{N}_\Gamma^0 \cong \{a \in \Omega^1(M^3; \mathfrak{su}(2)) \text{ such that } d_\Gamma * a = 0, ||a||_{C^m} < \epsilon\} \times_{\mathrm{Stab}(\Gamma)} \mathrm{SU}(2).$$

Where $\mathrm{Stab}(\Gamma) \subset \mathrm{SO}(3)$ is the stabilizer of Γ.

However, such a decomposition only holds on an open set for any given t. In general we will have to make several "cuts" and cover the moduli space with open sets of the form $\mathcal{M}^0_{+,\sigma} \times_{\mathcal{N}^0_\sigma} \mathcal{M}^0_{-,\sigma}$.

We want our open covering to be as canonical and useful as possible. We therefore wish to find a good estimate for the number of cuts we need to make to cover the moduli space. The properties of the Chern-Simons functional give a second kind of quantization that will provide such an estimate.

We introduce the notation $\{t_1, t_2\}$ for $[t_1, t_2] \times M^3$ and $\{t\}$ for $\{t\} \times M^3$. We pick a flat connection Γ on M^3 as a reference connection and note that every connection on $\{t_1, t_2\}$ is of the form

$$A = \Gamma + a + \phi dt$$

where $i_{\partial_t} a = 0$. We think of a and ϕ as a "time dependent" $\mathfrak{su}(2)$-valued 1-form and 0-form respectively. In fact there is a gauge transformation of A such that $\phi = 0$; this is the so-called "temporal gauge" and we will use this gauge to think of $\Gamma + a$ as a connection on M^3 with t as a parameter. The Chern-Simons functional arises from considering the energy of connections on a tube:

$$\begin{aligned}
\int_{\{t_1,t_2\}} \operatorname{tr}(F_A \wedge F_A) &= \int_{\{t_1,t_2\}} \operatorname{tr}(d_\Gamma a \wedge d_\Gamma a + 2 d_\Gamma a \wedge a \wedge a) \\
&= \int_{\{t_1,t_2\}} d\operatorname{tr}(a \wedge d_\Gamma a + \tfrac{2}{3} a \wedge a \wedge a) \\
&= \int_{\partial\{t_1,t_2\}} \operatorname{tr}(a \wedge d_\Gamma a + \tfrac{2}{3} a \wedge a \wedge a) \\
&= \int_{\{t_1\}} \operatorname{tr}(a \wedge d_\Gamma a + \tfrac{2}{3} a \wedge a \wedge a) - \int_{\{t_2\}} \operatorname{tr}(a \wedge d_\Gamma a + \tfrac{2}{3} a \wedge a \wedge a) \\
&\equiv CS(\Gamma + a_{t_1}) - CS(\Gamma + a_{t_2}).
\end{aligned}$$

Claim 2.2.2. The Chern-Simons functional $CS : \mathcal{A}(P|_{M^3}) \to \mathbb{R}$ defined as above satisfies the properties

1. $CS(g^*(A)) = CS(A) + \text{constant} \cdot \deg(g)$ where $g : M \to SO(3)$ is a gauge transformation on $P|_{M^3}$.

2. CS is constant on the components of ρ_{M^3}.

Proof. We calculate the differential of CS. Recall that $\mathcal{A}(P|_{M^3})$ is an affine space modeled on $\Omega^1(M^3, \mathfrak{su}(2))$. We evaluate the differential on a vector a at the connection A:

$$\begin{aligned}
dCS_A(a) &= \frac{d}{dt}(CS(A + ta))|_{t=0} \\
&= \int \operatorname{tr}(a \wedge dA + A \wedge da + 2a \wedge A \wedge A) \\
&= \int \operatorname{tr}(a \wedge F_A + A \wedge d_A a) \\
&= 2\int \operatorname{tr}(F_A \wedge a).
\end{aligned}$$

Assertion (2) is then immediate. From Stokes' theorem and the Bianchi identity it follows that dCS annihilates infinitesimal gauge transformations $d_A\xi$. The expression $CS(A^g) - CS(A)$ can thus only depend on the component of $\mathcal{G}(P|_{M^3})$ that g lies in, which is determined uniquely by the degree of g.

The Chern-Simons functional quantizes the amount of energy that an ASD connection can accumulate on a tube between two strips of small curvature. The representation variety ρ_M of flat connections on M is compact and real-algebraic (it is a quotient of a compact subset of $\mathrm{SU}(2)^N$ by a compact group) and so has a finite number of components. There is therefore a minimal non-zero number that occur as the difference $CS(\Gamma) - CS(\Gamma')$ for Γ and Γ' in different components of ρ_M. For many M^3 including the case we are primarily interested in—circle bundles over Riemann surfaces—a stronger form of quantization is true:

Theorem 2.2.3. There is a constant $e = e_M$ depending only on the topology of M^3 with the following property. For any fixed $\epsilon > 0$ there is a $\delta > 0$ such that for every ASD connection A on $\{-L - \frac{1}{2}, L + \frac{1}{2}\}$ such that

$$\int_{\{-L-\frac{1}{2},-L+\frac{1}{2}\}} |F_A|^2 + \int_{\{L-\frac{1}{2},L+\frac{1}{2}\}} |F_A|^2 < \delta$$

there is a non-negative integer $\hat{E} \in \mathbb{Z}_{\geq 0}$ satisfying

$$\left| \int_{\{-L-\frac{1}{2},L+\frac{1}{2}\}} |F_A|^2 - \hat{E} \cdot e \right| < \epsilon.$$

Proof. See [**Ta3**] Proposition 4.5 for the case of non-trivial circle bundles over a Riemann surface.

For the rest of the notes we will assume that M is such that theorem 2.2.3 holds although all the arguments can be made to go through with only minor modifications.

We now recount the total energy on X in units of e. Define n by the equation

$$\mathcal{E} = \int_X |F_A|^2 = n \cdot e.$$

We can construct a metric g_L on X so that we can cover the energy $\mathcal{E} = n \cdot e$ moduli space $\mathcal{M}^0_X$ by $n+1$ open sets. Let δ be as in the theorem above, and choose l satisfying $l\delta > n$. Fix the metric on X to be g_{nl}, the metric with n tubes of length l inserted. For notational purposes, we will take the length coordinate on the tube to run from 0 to nl.

Lemma 2.2.4. *Given $[A] \in \mathcal{M}_X$ there exists $m \in \{0, \dots, n\}$ such that*

$$\int_{\{lm-\frac{1}{2},lm+\frac{1}{2}\}} |F_A|^2 < \delta$$

and so that if $X_\pm$ denote the pieces of X cut along $\{ml\}$, the total energy of A on X_- and X_+ is ϵ-close to ke and $(n-k)e$ respectively.

Proof. Define the set

$$R = \{s : \int_{\{s-\frac{1}{2},s+\frac{1}{2}\}} |F_A|^2 < \delta\} \subset [0, lm].$$

R will be a collection of disjoint intervals $[s_1, s_1'], \dots, [s_q, s_q']$. Assume that the ASD connection A is in violation of Lemma 2.2.4. The contradiction arises by considering the possible distribution of the intervals $[s_i, s_i']$ in conjunction with theorem 2.2.3.

The upshot is that then we can cover the energy $n \cdot e$ moduli space:

$$\mathcal{M}_X^0 = \bigcup_{k \in \{0, \dots, n\}} \mathcal{M}_{X,k}^0$$

where $\mathcal{M}_{X,k}^0$ are open sets diffeomorphic to a fibered product of ASD moduli spaces on $X_- \cup \{0, lk\}$ and $\{lk, ln\} \cup X_+$ of energy approximately ke and $(n-k)e$ respectively, fibered over a regular neighborhood of the flat connection, *i.e.*

$$\mathcal{M}_{X,k}^0 = \mathcal{M}_{-,k}^0 \times_{\mathcal{N}^0} \mathcal{M}_{+,n-k}^0.$$

LECTURE 3
Simplicial Decomposition of $\mathcal{M}_X^0$

3.1. The decomposition

Recall that in the last lecture we used the properties of the Chern-Simons functional to determine a "quantum" of energy $e = e_M$. For the M^3 that we consider, the difference of the values that the Chern-Simons functional can take on different flat connections are integral multiples of e_M. We used e_M as units for counting the energy of ASD connections. For a fixed bundle P we fixed $n \in \mathbb{N}$ by $\int_X |F_A|^2 = n \cdot e$.

We then fixed a metric on X with a subset isometric to $\{0, nl\} \equiv [0, nl] \times M^3$ where l is a large fixed number determined so that for any A there is a $m \in \{0, \ldots, n\}$ such that A has small energy in a strip $\{ml - \frac{1}{2}, ml + \frac{1}{2}\}$ where small is in regard to some fixed small number ϵ.

We wish to use this fact to cover the moduli space with open sets that are isomorphic to fibered products and at the same time we want to keep the overlaps as simple as possible.

We would cover the moduli space with closed sets that are diffeomorphic to manifolds with corners, so that the overlaps only occur on codimension 1 faces. In terms of integrating differential forms this is more natural and easier to work with; however, for technical reasons, in [**Ta3**] the open cover must be used in conjunction with cutoff functions. As far as the structure of the overall results it makes no difference, so for simplicity we will define a decomposition into manifolds with corners.

We can define an open covering by

$$\mathcal{M}_X = \bigcup_{k=0}^{n} \mathcal{M}'_{X,k}{}^{1}$$

with $\mathcal{M}'_{X,k}$ defined to be the set of equivalence classes $[A] \in \mathcal{M}_X$ that satisfy

1. $\int_{\{lk-\frac{1}{2}, lk+\frac{1}{2}\}} |F_A|^2 < \delta$
2. $\int_{X_+ \cup \{lk, ln\}} |F_A|^2 \sim (n-k) \cdot e$.

---

[1]Recall that we work with the based moduli spaces so that we can express the sets as fiber products. However, in this lecture we will drop the ^{0} to avoid excessive notation.

Here we've introduced the notation $\sim$ to denote "ϵ-close", *i.e.* we say $a \sim b$ if and only if $|a-b| < \epsilon$.

We wish to refine the sets $\mathcal{M}'_{X,k}$ to closed manifolds with corners $\mathcal{M}_{X,k}$ that only overlap on codimension 1 faces. We define $\mathcal{M}_{X,k} \subset \mathcal{M}'_{X,k}$ by the conditions:

1. Same as (1) above.

2. Same as (2) above.

3. $\int_{\{(k-\frac{1}{2})l,kl\}} |F_A|^2 \leq \frac{1}{100}e$

4. If there is a $k' > k$ with $\int_{\{k'-\frac{1}{2},k'+\frac{1}{2}\}} |F_A|^2 < \delta$ then
$$\int_{\{lk,lk'\}} |F_A|^2 \leq (k'-k)e.$$

5. If the inequality in (4) is equality, then
$$\int_{\{(k'-\frac{1}{2})l,k'l\}} |F_A|^2 \geq \frac{1}{100}e.$$

A consequence of these conditions is:

Lemma 3.1.1. *The sets $\mathcal{M}_{X,k}$ cover $\mathcal{M}_X$ and if the inequalities in (3) and (5) are strict, then $[A]$ belongs to a unique set.*

We have the decomposition
$$\mathcal{M}_{X,k} = \mathcal{M}_{-,k} \times_{\mathcal{N}} \mathcal{M}_{+,n-k}$$
where $\mathcal{M}_{+,n-k}$ is defined to be equivalence classes of ASD connections A on $\{kl-\frac{1}{2}, nl\} \cup X_+$ satisfying conditions (1), (2), (4), and (5) above and $\mathcal{M}_{-,k}$ is equivalence classes of $[A]$ on $X_- \cup \{0, lk+\frac{1}{2}\}$ satisfying (1) and (3) and the analogous statement to (2):
$$\int_{X_- \cup \{0,kl\}} |F_A|^2 \sim k \cdot e.$$

The boundary of these sets can also be expressed as unions of fibered products. We use $\sqcup$ to denote a union of sets with intersection of codimension > 0 and summarize our decomposition as follows:

$$\begin{aligned}
\mathcal{M}_X &= \bigsqcup_{k=0}^{n} \mathcal{M}_{-,k} \times_{\mathcal{N}} \mathcal{M}_{+,n-k} \\
\partial \mathcal{M}_{+,q} &= \bigsqcup_{p=1}^{q} X_p \times_{\mathcal{N}} \mathcal{M}_{+,q-p} \\
\partial \mathcal{M}_{-,q} &= \bigsqcup_{p=1}^{q} \mathcal{M}_{-,q-p} \times_{\mathcal{N}} X_p \\
\partial X_q &= \bigsqcup_{p=1}^{q-1} X_p \times_{\mathcal{N}} X_{q-p}
\end{aligned} \tag{3.1}$$

X_q is the moduli space of ASD connections on $\{-\frac{1}{2}, lq+\frac{1}{2}\}$ satisfying the conditions:

1. $\int_{\{-\frac{1}{2},+\frac{1}{2}\}} |F_A|^2 < \delta$ and $\int_{\{lq-\frac{1}{2},lq+\frac{1}{2}\}} |F_A|^2 < \delta$.
2. $\int_{\{0,lq\}} |F_A|^2 \sim q \cdot e$.
3. $\int_{l(q-\frac{1}{2}),lq} |F_A|^2 = \frac{1}{100}e$, (note that this is a *closed* condition).
4. If $q' < q$ and $\int_{\{lq'-\frac{1}{2},lq'+\frac{1}{2}\}} |F_A|^2 < \delta$ then
$$\int_{\{0,lq'\}} |F_A|^2 \leq q' \cdot e \text{ and}$$
5. if the inequality in (4) is equality then
$$\int_{\{l(p'-\frac{1}{2}),lp'\}} |F_A|^2 \geq \tfrac{1}{100}e.$$

Restriction of A on the slices $\{0\}$ and $\{lp\}$ give two maps j_- and j_+ respectively each mapping X_q to $\mathcal{N}$. When X_q is the left (right) factor of a fibered product it is understood that the map is j_+ (j_-).

We leave the verification of equations 3.1 to the reader. It is essentially bookkeeping using the definitions and the quantization principle for energy of ASD connection on the tube.

3.2. Formal consequences

Recall that to compute Donaldson invariants we wish to evaluate an integral on $\mathcal{M}_X$. For the moment we will pretend that all our fibered products satisfy the hypothesis of theorem 2.1.1 and see the formal consequences:

$$\begin{aligned}\int_{\mathcal{M}_X} \omega_- \wedge \omega_+ &= \sum_{k=0}^{n} \int_{\mathcal{M}_{-,k} \times_{\mathcal{N}} \mathcal{M}_{+,n-k}} \omega_- \wedge \omega_+ \\ &= \sum_{k=0}^{n} \int_{\mathcal{M}_{+,n-k}} i_+^*(i_{-*}\omega_-) \wedge \omega_+.\end{aligned}$$

We would like to define $\sum_{k=0}^{n}(i_{-*}\omega_-)_{\mathcal{M}_{-,k}}$ to be a relative Donaldson invariant. In order to find a Floer cohomology theory to be the receptacle for our candidate invariant we need to answer the question:

- What differential D annihilates $\sum_{k=0}^{n}(i_{-*}\omega_-)_{\mathcal{M}_{-,k}}$ when $d\omega_- = 0$?

Likewise, to find a Floer homology theory we should answer the question:

- If $d\omega_+ = 0$, what D exists so that for all μ
$$\sum_{k=0}^{n} \int_{\mathcal{M}_{+,k}} i_+^*(D\mu) \wedge \omega_+ = 0?$$

D should be a differential on the algebra of distribution-valued differential forms on $\mathcal{N}$.

Theorem 3.2.1 (Formal). *Define D by the equation*

$$D = d + \sum_{k\geq 1} j_{-*}(j_+^*(\cdot))_{X_k},$$

then

- $i_{-*}(d\alpha_-) = D(i_{-*}\alpha_-)$ *when α_- is a form that "comes from X_-",*
- $D^2 = 0$, *and*
- $\int_{\mathcal{M}_+} i_+^*(\mu) \wedge d\alpha_+ = \int_{\mathcal{M}_+} i_+^*(D\mu) \wedge \alpha_+$ *when α_+ is a form that "comes from X_+".*

Proof. We will use the following version of Stokes' theorem: If $f : Y \to Z$ is a proper Fredholm map and Y and Z are manifolds with boundary, then

$$f_* d = df_* + (f_*)_{\partial Y}.$$

The first claim follows from the boundary relations for the $\mathcal{M}_{-,k}$'s, the second from the boundary relations for the X_q's, and the last from the boundary relations for the $\mathcal{M}_{+,k}$'s. The proofs are straight forward but notational tedious. One needs to use the obvious compatibility conditions for the maps $i_\pm$ and $j_\pm$ that are not explicitly stated in equations 3.1. For example the restriction of $i_- : \mathcal{M}_{-,k} \to Z$ to the boundary component $\mathcal{M}_{-,k-q} \times_Z X_q$ is $j_+ \circ \pi_{X_q}$ (whereas the map defining the fibered product is j_-). The "coming from X_-" conditions are that α_- restricted to $\mathcal{M}_{-,k} \times_Z \mathcal{M}_{+,n-k}$ is of the form $\pi^*_{\mathcal{M}_{-,k}}(\alpha_{-,k})$ and $\alpha_{-,k}$ restricted to the boundary component $\mathcal{M}_{-,q} \times_Z X_{k-q}$ is $\pi^*_{\mathcal{M}_{-,q}}(\alpha_{-,q})$ and similarly for α_+.

Theorem 3.2.1 implies that the assignment $\omega_- \mapsto i_{-*}(\omega_-)$ defines a homomorphism in cohomology:

$$\mathcal{D}_{X_-} : H_d(\mathcal{M}_-) \to H_D(\mathcal{N}).$$

Also

$$\mathcal{D}_{X_+} : H_d(\mathcal{M}_+) \to \mathrm{Hom}(H_D(\mathcal{N}); \mathbb{R})$$

is induced by

$$\omega_+ \mapsto \int_{\mathcal{M}_+} i_+^*(\cdot) \wedge \omega_+.$$

So at least formally, we have achieved our goal. The Donaldson invariants of X have been expressed in terms of relative invariants of $X_\pm$ that take values in a Floer (co)homology. We have that

$$\int_{\mathcal{M}_X} \omega_- \wedge \omega_+ = \langle \mathcal{D}_{X_-}(\omega_-), \mathcal{D}_{X_+}(\omega_+) \rangle$$

and

$$HF^*(M^3) = H_D(\mathcal{N}).$$

3.3. Problems revisited

The preceding discussion is purely formal until we address the following three issues:

1. The maps $i_\pm : \mathcal{M}_\pm \to \mathcal{N}$ are not Fredholm so
    - the push forward i_{-*} is not well defined and

- the integral $\int_{\mathcal{M}_+} i_+^*(i_{-*}\omega_-)\wedge\omega_+$ is over an infinite dimensional space.

2. We really want to integrate over the unbased moduli spaces.

3. Even when we fix problem (1), there will still be non-compactness due to bubbling off.

Recall that to deal with issue (2) we introduced a form α on the based moduli space that is the volume form restricted to each SO(3)-orbit and is invariant under the action. We can then pull back the integrand to the based moduli space and integrate it against α. Theorem 3.2.1 as stated fails with the presence of α; a more complicated definition of D is required that incorporates the group action on $\mathcal{N}$. This leads to "equivariant" versions of Floer homology that in practice arise when there are reducible flat connections on M^3 or the intersection form of X_+ or X_- is negative definite (see [**Ta2**], [**Br**]). The complications are essentially algebraic in nature and will not be considered in these notes.

Issue (3) is a technical issue that plagues most every aspect of Donaldson theory. The difficulty is always that the manifolds that we integrate over have non-compactness due to bubbling off. The rough idea is to compactify these spaces as manifolds with corners and then to restrict the complex of differential forms to a subcomplex that behave nicely restricted to the boundary. See [**Ta3**] chapter 5, appendix A for the gory details.

Issue (1) is a central one and will be the primary focus of the subsequent lecture.

LECTURE 4
Half-Infinite Dimensional Spaces

Recall that in the last lecture we formally constructed relative Donaldson invariants and Floer homology using a decomposition of the moduli space into fibered product spaces. The relevant maps in the fibered products are

$$\begin{aligned} i_\pm : &\quad \mathcal{M}_\pm \to \mathcal{N} \\ j_\pm : &\quad X \to \mathcal{N} \end{aligned}$$

where $\mathcal{M}_\pm$ and X are moduli spaces of ASD connections on $X_\pm$ and $[-L, L] \times M^3$ respectively and $\mathcal{N}$ is a regular neighborhood of the flat connections in $\mathcal{B}_{M^3}$. The metric on $X_\pm$ and the length L depend on the energy of the corresponding ASD connection (it varies from component to component in the moduli spaces).

The constructs of the previous lecture were made under the false assumption that $i_\pm$ and $j_\pm$ are Fredholm maps. In this section we show how we can modify $\{\mathcal{M}_\pm, X, \mathcal{N}, i_\pm, j_\pm\}$ so that the arguments of section 3.2 apply.

The map $i_- : \mathcal{M}_- \to \mathcal{N}$ is morally an embedding of infinite dimension and codimension. We want to find a projection π of $\mathcal{N}$ onto a "half-infinite dimensional" subspace $\mathcal{Z} \subset \mathcal{N}$ so that the composition of i_- with π is Fredholm.

$\mathcal{Z}$ can be constructed using the techniques of [**Ta3**] and [**Ta5**] even when the variety of flat connections has singularities. Let us assume here, though, that the flat connections form a smooth manifold. $\mathcal{N}$ is a normal neighborhood of the flat connections and let $\mathcal{N}_\Gamma \subset \mathcal{N}$ be the fiber over Γ of a projection from $\mathcal{N}$ to the flats. We will construct half-infinite dimensional spaces $\mathcal{Z}_\Gamma \subset \mathcal{N}_\Gamma$ fiberwise. There is an SU(2)-equivariant diffeomorphism:

$$\mathcal{N}_\Gamma \cong \left\{a \in \Omega^1(M^3; \mathfrak{su}(2)) : d_\Gamma *_M a = 0\right\} \times \mathrm{SU}(2)$$

Define

$$K_\Gamma = \left\{a \in \Omega^1(M^3; \mathfrak{su}(2)) : d_\Gamma *_M a = 0\right\}.$$

Completing K_Γ in an appropriate Sobolev norm we have that $*d_\Gamma : K_\Gamma \to K_\Gamma$ is a symmetric operator with isolated real eigenvalues that are unbounded from above and below. We have

$$K_\Gamma = \bigoplus_{\lambda \in \mathrm{Spec}(*d_\Gamma)} K_{\Gamma,\lambda}$$

where $K_{\Gamma,\lambda}$ are the (finite dimensional) eigenspaces of $*d_\Gamma$. Γ is called *non-degenerate* if $0 \notin \text{Spec}(*d_\Gamma)$. We have been assuming that all the flat connections are non-degenerate (although the machinery of [**Ta3**] handles the general case). We then have the decomposition into positive and negative eigenspaces $K_\Gamma = K_{\Gamma,-} \oplus K_{\Gamma,+}$. Now let

$$\mathcal{Z}'_\Gamma = K_{\Gamma,-} \times \text{SU}(2) \subset \mathcal{N}_\Gamma.$$

$\mathcal{Z}_\Gamma$ will be a small deformation of $\mathcal{Z}'_\Gamma$ that is tangent to $\mathcal{Z}'_\Gamma$ at Γ. $\mathcal{Z}_\Gamma$ can be constructed as follows. Define $\mathcal{M}_{\Gamma,-}$ to be equivalence classes of ASD connections on $(-\infty, 0] \times M^3$ satisfying

1. $\int_{\{-\infty,0\}} |F_A|^2 < \nu$ where ν is a fixed small positive number, and
2. $\lim_{t\to-\infty}[A]_t \to [\Gamma]$.

We define $\mathcal{M}_{\Gamma,+}$ in a similar way, using the cylinder $\{0,\infty\}$. The map $\mathcal{M}_{\Gamma,-} \to \mathcal{N}_\Gamma$ induced by restriction to $\{0\}$ is a "half-infinite dimensional" embedding and is a candidate for $\mathcal{Z}_\Gamma$.

Why is the image of $\mathcal{M}_{\Gamma,-}$ a small perturbation of $\mathcal{Z}'_\Gamma$? On the cylinder $\{-\infty, 0\}$ the ASD equations for a connection $\Gamma + a$ near Γ to a first order approximation are

$$\frac{da}{dt} + *d_\Gamma a = 0.$$

We see that the eigenvalues of $*d_\Gamma$ control the behavior of the solutions. In order to have bounded solutions on $\{-\infty, 0\}$, a must be composed of eigenfunctions with non-positive eigenvalues. Connections in $\mathcal{M}_{\Gamma,-}$ can be obtained from solutions of the linear equations by small perturbations that take into account the non-linear terms.

Once we have $\mathcal{Z}_\Gamma \subset \mathcal{N}_\Gamma$ we choose a projection $\pi_\Gamma : \mathcal{N}_\Gamma \to \mathcal{Z}_\Gamma$. For example, there exists a projection such that $\pi_\Gamma^{-1}(\Gamma) = \mathcal{M}_{\Gamma,+}$. For circle bundles over Riemann surfaces these constructions can be made so that they piece together to form a smooth manifold $\mathcal{Z}$ and a projection $\pi : \mathcal{N} \to \mathcal{Z}$ that satisfy:

Theorem 4.0.1. *$\pi \circ i_- : \mathcal{M}_- \to \mathcal{Z}$ is a Fredholm map.*

Thus if $g : N \to \mathcal{Z}$ is a map of a finite dimensional manifold to $\mathcal{Z}$ that is transverse to $\pi \circ i_-$ then $\mathcal{M}_- \times_{\mathcal{Z}} N$ is a union of finite dimensional manifolds. In particular, with the choice of π described above, if ρ_σ is a smooth component of the flat connections then $(\pi \circ i_-)^{-1}(\rho_\sigma)$ is the L^2-moduli space of ASD connections on $X_- \cup \{0, \infty\}$ that have limiting values in ρ_σ.

From the X_+ side we have:

Theorem 4.0.2. Let $f_+ = \pi \circ i_+$. df_+ is a compact operator; furthermore, the projection of di_+ to $\text{Ker}\, d\pi$ is Fredholm.

Morally this is saying that $i_+^{-1}(\mathcal{Z})$ is a finite dimensional real analytic variety.

If the fibered products of section 3.2 could be made to be over $\mathcal{Z}$ then we could hope that the hypotheses to 2.1.1 hold. To achieve this one needs to perturb the spaces $\{\mathcal{M}_-, \mathcal{M}_+, X\}$ to spaces $\{\hat{\mathcal{M}}_-, \hat{\mathcal{M}}_+, \hat{X}\}$ that have a decomposition analogous to 3.1 but with the fibered products over $\mathcal{Z}$. Such perturbations are possible for a large class of M^3 including homology spheres with non-degenerate flat connections, circle bundles over Riemann surfaces with negative first Chern class, and $S^2 \times S^1$.

How is such a perturbation achieved? It is a deformation of the boundary conditions. For example $\hat{X}_q$ is defined to be the collection of equivalence classes of connections $\{[A_1], \ldots, [A_q]\}$ satisfying:

1. A_i is ASD on $\{(i-1)l, il\}$.
2. $\sum_{i=1}^{q} \int_{\{(i-1)l,il\}} |F_{A_i}|^2 \sim q \cdot e$.
3. $\int_{\{0,\frac{1}{2}\}} |F_{A_1}|^2 < \delta$ and $\int_{\{ql-\frac{1}{2},ql\}} |F_{A_q}|^2 < \delta$.
4. $\int_{\{(q-\frac{1}{2})l,ql\}} |F_{A_q}|^2 = \frac{1}{100}e$.
5. $i_-[A_1] \in \mathcal{Z}$.
6. If $p \in \{1, \ldots, q-1\}$, $\int_{\{pl,pl-\frac{1}{2}\}} |F_{A_p}|^2 < \delta$, and $\int_{\{pl,pl+\frac{1}{2}\}} |F_{A_{p+1}}|^2 < \delta$, then
$$\sum_{i=1}^{p} \int_{\{(i-1)l,il\}} |F_{A_i}|^2 \leq p.$$
7. If the inequality in (6) is equality then:
   - $\int_{\{(p-\frac{1}{2})l,pl\}} |F_{A_p}|^2 \geq \frac{1}{100}e$,
   - $\pi \circ i_+[A_p] = \pi \circ i_-[A_{p+1}]$, and
   - $$\beta\left(\frac{100}{e}\int_{\{(p-\frac{1}{2})l,pl\}} |F_{A_p}|^2\right) \cdot v \circ i_+[A_p] = v \circ i_-[A_{p+1}]$$
   where β is a smooth cutoff function such that $\beta(t)$ is 0 for $t < 1$ and is 1 for $t > 2$ and v is projection onto $\operatorname{Ker} d_\Gamma^*$.
8. Finally, if $p \in \{1, \ldots, q-1\}$ and (6) does not hold then $i_+[A_p] = i_-[A_p]$.

With the preceding definition, and letting $r_+ = \pi \circ i_+$ and $r_- = i_-$. We also reindex X_q as X_m so that $\operatorname{Index}(dr_+|_{X_m}) = m$ (this is a finite to one reindexing) . We then have:

Theorem 4.0.3. *The operator*
$$D = d + \sum_{m \geq 0} r_{+*}(r_-^*(\cdot))_{\hat{X}_m}$$
is a well-defined operator on distribution-valued forms on $\mathcal{Z}$ and it satisfies $D^2 = 0$. Furthermore, we can fix a $\pmod 8$ *grading on $\Omega^*(\mathcal{Z})$ so that D increases the degree by* 1.

Here the grading on a differential form on a given component of $\mathcal{Z}$ is defined in terms of the form degree and the spectral flow of the component.

We will conclude by briefly remarking that the grading on $\Omega^*(\mathcal{Z})$ can be arranged so that there is a natural filtration and thus, a spectral sequence. The spectral sequence converges to the D cohomology of $\Omega^*(\mathcal{Z})$. When the sequence is applied to the case of a homology sphere with non-degenerate flat connections, Floer's original picture of instanton homology is recovered. See also [**Au-Br**].

References

[At] M. F. Atiyah *New Invariants for 3 and 4 Dimensional Manifolds.* Proc. Sympos. Pure Math. 48(1988).

[Au-Br] D. Austin, P. Braam *Equivariant Floer (Co)homology and Gluing Donaldson Polynomials*, preprint.

[Br-Do] P.J. Braam and S. K. Donaldson, *Fukaya-Floer Homology and Gluing Formulae for Polynomial Invariants.* Preprint.

[Br] J. Bryan *Symplectic Geometry and the Relative Donaldson Invariants of* $\overline{\mathbb{CP}}^2$. Preprint.

[Do-Kr] S. K. Donaldson and P. Kronheimer, *The Geometry of Four-Manifolds.* Oxford Science Publications.

[Fl] A. Floer, *An Instanton invariant for 3-manifolds* , Comm. in Math. Phys., 118 (1988), 215-240.

[Fr-Mo] Robert Friedman and John Morgan. *The Differential Topology of Complex Surfaces.* Springer-Verlag, 1994.

[Ta1] Clifford H. Taubes, *A Simplicial Model for Donaldson-Floer Theory.* Preprint.

[Ta2] Clifford H. Taubes, *The Role of Reducibles in Donaldson-Floer Theory.* To appear in the Proceedings of the 1993 Taniguchi Symposium on Low Dimensional Topology and Topological Field Theory.

[Ta3] Clifford H. Taubes, *Floer theory for Twisted Circle Bundles.* Preprint.

[Ta4] Clifford H. Taubes, unpublished

[Ta5] Clifford H. Taubes, L^2 *Moduli Spaces on 4-manifolds with Cylindrical Ends.* Monographs in Geometry and Topology, Vol. 1; International Press.